Many people perceive this to be a moment of despair. Global warming has reached terrifying heights of severity, human expansion has caused the extinction of countless species and neoliberalism has led to a destructive divide in wealth and a polarization of mainstream politics. But, there are constructive ways to meet these challenges, and there are plenty of reasons for optimism.

Lily Cole has met with some of the millions of people around the world who are working on solutions to our biggest challenges and are committed to creating a more sustainable and peaceful future for humanity. Exploring issues from fast fashion to fast food and renewable energy to gender equality, and embracing debate, the book features interviews with diverse voices from entrepreneurs Stella McCartney and Elon Musk, to activists Extinction Rebellion co-founder Dr Gail Bradbrook and Farhana Yamin, to offer a beacon of possibility in challenging times.

Who Cares Wins is a rousing call to action that will leave you feeling hopeful that we can make a difference. We are the ancestors of our future: a generation that will either be celebrated for its activism or blamed for its apathy. It is for us to choose optimism, to make changes and to create the future we want.

Who Cares Wins

Who Cares Wins

*Reasons for Optimism in
Our Changing World*

LILY COLE

PENGUIN LIFE

AN IMPRINT OF

PENGUIN BOOKS

PENGUIN LIFE

UK | USA | Canada | Ireland | Australia
India | New Zealand | South Africa

Penguin Life is part of the Penguin Random House group of companies
whose addresses can be found at global.penguinrandomhouse.com.

First published 2020
001

Copyright © Lily Cole, 2020

The moral right of the author has been asserted

Typeset in 12.8/16 pt Garamond MT Std
by Integra Software Services Pvt. Ltd, Pondicherry
Printed and bound in Great Britain by Clays Ltd, Elcograf S.p.A.

A CIP catalogue record for this book is available from the British Library

HARDBACK ISBN: 978–0–241–30912–4
TRADE PAPERBACK ISBN: 978–0–241–37175–6

www.greenpenguin.co.uk

Penguin Random House is committed to a
sustainable future for our business, our readers
and our planet. This book is made from Forest
Stewardship Council® certified paper.

For Wylde and wilderness

Contents

Preface

In the slim gap between my signing off the final text of this book affirming 'Reasons for Optimism in Our Changing World', and ink hitting the pages, our world changed, irrevocably. Or rather, the human world changed: the elemental world was changing a little less than usual. Who knew our machine could stop?

<u>Corona</u> (Covid-19) has offered a shock to the global psychology unlike anything experienced in modern times. A collective arrest in the rehab of our homes, to explore the prisons or palaces of our minds. Whilst we huddle together (two metres apart if privileged enough to social distance) in fear, confusion, love, and hope, many voices are quickly proclaiming this is the wake-up call the world desperately needed. The pattern interruption, the stillness: a chance to reflect on our habits.

Etymology: *corōna* (Latin) – 'garland worn on the head as a mark of honour or emblem of majesty, halo around a celestial body'; the name given to a group of microscopic viruses, fringed by spikes, thought to have passed to humans because of our unbalanced interactions with animals and wild spaces.

What will come next? When industries wake from their slumber, travel bans are lifted, when we can dance with strangers again – what will we have learned? Will we have made enough of a shift in how we think, consume and live, to do things differently – to avert another crisis?

Disasters have always brought with them pivots along our historical path. The charged beauty of this moment is that almost anything is possible: both Orwellian and utopian narratives haunt us. Whilst

Naomi Klein has analysed how the shock of disasters has been misused to instate authoritarian regimes, conversely crises have often led to very progressive changes: many national health services working hard today were a result of the 1918 flu epidemic; the First World War helped emancipate women; the Second led to the founding of the United Nations and modern welfare states.

Corona has revealed our capacity for collective mobilization, kindness and how quickly everything can be transformed. More has changed in a matter of weeks than many of us thought possible in years: and more will change again.

Might this experience enable us to fundamentally realize how connected we are with each other and the Earth? Could it wake us up to the huge financial and social costs of an exploitative relationship with nature? Might governments use this opening to make bold, positive policy shifts? What types of business do we want to survive? How will our values be transformed?

In the introduction to this book I recall the fable of the frog in water – as metaphor for our response to the climate crisis. The story went that a frog would quickly jump out of boiling water, but would stay in water that imperceptibly warmed, until its end.

The climate crisis emerges gradually. Too quickly for many of the Earth's species to adapt to, but too slowly for our imaginations to track. Seven million people dying annually of air pollution (nearly 20,000 a day)* would be shocking if it happened suddenly – but it has been a slow boil to get to that number, and it creeps ever higher.

Corona poured boiling water upon us, and we jumped. Will we return to our collective trance? Or can we use this pause to consider new ways of being? I hope this book might offer a voice in the conversation we urgently have to have, and some seeds of ideas we can choose to help blossom in a new – and healthier – reality.

---

* https://www.who.int/health-topics/air-pollution#tab=tab_1

OMG

Twenty-four icebergs were melting into the pavement in London.

Over several days, the icebergs glistened and contorted, softly sculpted by the warming air, which worked like Bernini's fingers upon them. Holes and gaps emerged and disappeared. Streaks of iridescent blue revealed the layered snowfall of tens of thousands of years.

Olafur Eliasson had brought them from the sea near Greenland for *Ice Watch* – an artistic reminder that 10,000 of these icebergs fall into the ocean every second.

I picked my daughter, Wylde, up early from nursery so she could see the ice before it ran into the River Thames.

Her instinctive impulse was to lick it and laugh.

OMG – Oceans Melting Greenland, NASA project measuring accelerating ice loss.

~~~

It's a strange time to be alive. Perhaps all times feel like peculiar ones, yet now feels particularly riddled with contradiction.

Is this the best moment ever?

Are we on the brink of extinction?

Can both be true?

From ancient religious texts, to medieval paintings and Hollywood movies, apocalypse has long haunted our imaginations, like a burning shadow.

Yet it is not just the artists, mystics or zealots who are raising the alarm.

This time, it's the scientists.

We are, they tell us – *as we party, love, queue, pay our taxes* – at the beginning of the Sixth Mass Extinction.

~~~

We may not be the first to dance into our death. In fact, many believe affluence and apocalypse travel together.

Easter Island, in the fourteenth to fifteenth centuries, was probably a good time and place to be alive. There were about fifteen thousand Rapa Nui indigenous people on the island. An ambitious, complex society, they carved elegant sculptures from volcanic ash, some ten metres tall. Nine hundred enigmatic ancestral faces, imbued with <u>*mana*</u>, continue to watch the world.

The sacred spirit of the Rapa Nui's ancestors was believed to live on in the 'Moai' sculptures.

Yet the Rapa Nui's industriousness, combined with over-population, and the introduction of the

2

Polynesian rat, led the island to become heavily defor-
ested. Land birds became extinct; the soil was eroded
and water resources dried up. When European colonizers arrived in 1722, the population had dwindled
to a few thousand people.

~~~

Infamously, if you put a frog in boiling water it will
jump out, but if you put a frog in warm water and slowly
heat it to boiling point, the frog will stay until its end.

Our planet is slowly warming around us, literally
melting, and yet we haven't jumped. Are we frogs,
dazzled and distracted by the rainbows cast in our
glass of warming water; too comfortable or condi-
tioned to take the leap?

Would leaving feel lonely? Weird? Un-cool?

~~~

Imagine if the dinosaurs or the dodo had directed
their own swansongs. We may become the first species to document our own extinction.

People make plays, songs and paintings about the
crisis. Look, I'm writing this, a book about the ways
in which people are trying to halt the Sixth Mass
Extinction.

Because another possibility exists –

We could become the first species to *prevent* a mass
extinction.

Dwell on that.

~~~

Solastalgia: the anguish caused by environmental alterations due to droughts and destructive mining. Taking the Latin word for comfort (sōlācium) and the Greek root designating pain (-algia): a neologism that sums up the devastating effects of finding unease where you used to look for relief.

I often feel like a frog.

To truly, deeply, accept what 99.99 per cent of the scientists are telling us is so profoundly earth-shattering, it makes it hard to carry on as normal, as life invariably encourages us to do. Whenever I do connect with the scientific facts, I find myself crying.

Then I ask myself, *how can we jump out?*

Trying to answer that question is the point of this book.

In many ways, I have written it for myself. You know how you can be so good at giving relationship advice to a friend but really, you're just talking to yourself?

~~~

If you dare to care, it's a head-fuck. The problem is complex, and the solutions often contradictory. One day, biofuels are heralded as our saviour: another day they are driving deforestation in Indonesia. Some say we should stop eating all meat; others say a small amount of holistic animal agriculture is part of the solution.

The aeroplane is the poster child of destruction; yet few people know it is often more energy efficient to fly than drive solo. It's hard to know which plastics are recyclable, harder still to know their afterlife when we put them in green bins.

How much do we need individual change; how much system change? Where does one end and the other begin? How powerful are we? How responsible are we?

Sometimes it feels like you can't even breathe without stealing oxygen.

~~~

Many years ago, I was outside in a field with a conservationist, admiring a compost toilet. It was a simple equation, he said – one day, we would have to come to terms with the fact that billions of people can't have flushable ones (or everything else they represent).

This tension seems to be at the heart of the contemporary environmental debate, and indeed this book:

Can we upgrade developed lifestyles to meet environmental demands?

Or do we need to change our lives entirely?

In the terminology of Charles Mann, 'Wizards' believe we can innovate our way out of the crisis, whereas 'Prophets' maintain that we need to radically simplify and cut back.

The Prophet's response is tempting: to withdraw, go off grid and get your nails dirty with earth. It seems to be the wisest response to the situation we find ourselves in, but is it pragmatic? Can everyone afford to live off the land? Will stepping back ever be mainstream? Is saying no to human innovation philosophically dubious?

Is my friend who built a cabin in a field in Cornwall, and works on a local organic farm, helping the planet more, say, than a technologist who flies around the world, spearheading an electric vehicle and solar energy revolution?

~~~

For those of us not living in the woods, how do we sit comfortably with imperfection? Is it better to have billions of imperfect environmentalists than thousands of perfect ones?

How dare I write about the environment when I am inevitably part of the problem?

This thought has tortured me to the point of paralysis. Hypocrisy seems inherent in writing a book like this, and I want to make clear at the outset, I am by no means perfect. It is a reality I continue to try to make peace with. So many of us are part of the problem. That's *the problem*. It is so pervasive, so embedded in all our normalized everyday choices, that it's very hard to escape it.

How much hypocrisy is *too much* hypocrisy?

~~~

I've lived an unusual and lucky life, which has brought me diverse experiences and eclectic friends. It's a

6

strange recipe, producing no doubt a strange cake, but I feel grateful for the broad perspective life has given me, bridging many different realities.

At a time of heightened sensitization and polarization – when talking and listening to each other feels more important than ever – I hope, in this book, to bring together some of the disparate voices and experiences, which I have been fortunate to have had access to.

In every topic this book touches on, a cacophony of (well-meaning) divergent perspectives springs alive. I have intentionally explored these, trying to retain an open mind, even – and especially – when they contradict each other. Occasionally an opinion of mine might sneak in, but I try to avoid or challenge them, because I am not interested in ideological fights, or claims to right and wrong.

I am interested in building bridges; in engaged, proactive optimism; in all the frogs jumping out of the water.

Why bother with optimism?

Being optimistic makes you happier, and is more likely to inspire joy in the people around you.

Whereas pessimism can breed apathy, inaction and fatalism – *if you believe the ship is sinking, there is no point fixing the hull* – optimism can inspire action and change. We will only radically innovate or even bother to sign petitions, if we believe that the future can be saved.

Optimism leads to positive behaviour change, which can make it a self-fulfilling prophecy. An example is health: optimists have been found to be more likely to exercise and follow medical advice, and less likely to smoke. According to a Harvard study, an optimistic outlook early in life can increase life expectancy by fifteen to forty years.

So, what would happen if we applied optimism to our social and environmental realities?

Could we write better prophecies to fulfil?

~~~

My first exposure to environmentalism came in the form of a yawn in the sky. We had been asked to choose a subject for a primary school project, and for some reason, I chose the ozone hole. I remember sketching an old-school fridge and people getting sunburnt.

The hole seemed like an ever-expanding, evanescent threat: an operatic gaping mouth that would eventually gobble us all up . . . but then, in 2018, NASA announced that the ozone hole was slowly healing.

The childhood monster in my mind died.

The Montreal Protocol, which had banned the production of ozone-depleting chemicals the year I was born (1987), had proven extremely successful.

Not that it was perfect. In 2016 the Montreal Protocol was amended in Kigali, Rwanda, by 170 countries, to commit to phase out HFCs, the primary replacement for CFCs, which spare the ozone but were found to be thousands of times more impactful than carbon dioxide as a greenhouse gas.

Scientists estimate the Kigali Accord will reduce global warming by nearly 1 degree Fahrenheit. The correct disposal of existing fridges and air-conditioner units has been ranked by climate research organization Project Drawdown as the *number one climate solution* that could have the very biggest reduction in emissions. (Though trumped by women's empowerment if you add together the impacts of women's education and family planning.)

The Montreal Protocol was the first universally ratified international treaty in the United Nations' history, and what Kofi Annan called 'perhaps the single most successful international agreement'.

The story of the ozone layer also shows the resilience of nature: that it can literally heal its own holes, when we act to protect it.

A map of the world that does not include Utopia is not worth even glancing at, for it leaves out the one country at which Humanity is always landing. And when Humanity lands there, it looks out, and, seeing a better country, sets sail. Progress is the realization of Utopias.
Oscar Wilde, *The Soul of Man Under Socialism*

Etymology: based on Greek *ou* 'not' + *topos* 'place'. Whilst some people like to think this means the good place is no place (i.e. it doesn't exist), others like to think it means that no place is the good place (i.e. utopia is an idea and sentiment, rather than a thing).

In the miasma of fear and negative news, we seldom stop to reflect on the narrative of progress that our age is a consequence of.

Humans have built domes and bridges, composed music, and developed complex artistic cultures. We have discovered electricity, deconstructed our DNA and travelled to the Moon. We have developed cures for diseases and doubled our lifespans. In the last few decades alone, maternal mortality has decreased by 44 per cent and infant mortality halved.

Many people live lives of material comfort far more luxurious than that of a king or queen even a hundred years ago: able to adjust the heat and lighting of their living environments, get clean drinking water from a tap, swallow magical pills whenever they feel sick, buy exotic food and products from all corners of the world, travel at great speed, and access libraries of information at their fingertips. Is this possible for everyone? Wizards think so.

Numerous peace treaties have been signed, and many people are living in the most peaceful human societies the planet has ever known. Slavery is officially illegal worldwide, and basic universal human rights have been established.

Of course, our problems have not been solved and our social systems are far from perfect, but the vector of change points in a positive direction, and a lot of people are working hard to solve the challenges that remain.

As our circumference of empathy widens, humanity has begun to establish rights for animals and the natural world. Through decades of tireless work by conservationists, campaigners, NGOs, scientists and inventors, marine and wildlife reserves have been extended, regulations have been put in place, and some species have been brought off the endangered species list.

This progress has not been enough, hence the apocalyptic warnings from the United Nations Intergovernmental Panel on Climate Change, calling for 'rapid and far-reaching' changes to be made by 2030.

The question arises: can we?

An optimist would say yes, and then set about trying to.

~~~

All over the world, children and adults are rising up, giving voice to the voiceless. In a very short space of time, environmentalism has gone from being a niche and largely academic concern, to a mainstream grass-roots movement, affecting corporate and political policy. Meanwhile, science and technology are innovating at an insanely rapid pace.

We may not be able to 'stop' global warming, but we can reduce the rate of escalation, and every per cent of a degree matters. We also need to think about adaptation, and how to minimize the social fallout of change: strong legal and democratic structures, compassionate migration policies, self-sufficiency, community, kindness.

We have many tools, and ideas to transform our systems, and ourselves.

This book is dedicated to some of them.

~~~

Environmental and social issues are, of course, deeply intertwined. I have tried to look at the problem from multiple perspectives, loosely charting my own journey these last fifteen years, as I have tried to understand my own role, responsibility and agency.

I begin small, thinking about the impact, and power, of our daily purchasing choices, then in Part Two I look at technology, the role it can play and our relationship to it. These sections may feel bourgeois – exploring the decisions that can be made by people who have the luxury of choice – but given that the richest 10 per cent of the world are responsible for half of all emissions, it is arguably a priority to understand and improve the impact that wealth affords.

The United Kingdom is particularly well represented in the upper echelons of global wealth, with 65 per cent of the population in the richest 10 per cent of people globally and a whopping 93 per cent of the population in the top 20 per cent.

We are citizens first and foremost, and in Part Three I look at some of the transformative bi-partisan political ideas on the table, and how we might help to see them happen. Finally, in Part Four, perhaps the most important, I explore what we might learn from completely different societies, offering alternative values and mindsets. We have a unique opportunity to marry modern innovation with indigenous wisdom.

~~~

I hope some of the ideas contained in this book will inspire and empower you. They are not meant to be prescriptive. I am not a specialist or a scientist, but I am curious and always trying to learn. These are the bastions of hope I cling to, and sometimes act upon, as I try to be optimistic, useful and ultimately happy.

The future predicted by statisticians is not written in stone, and what I want to propose in this book is that you have the power to shape it.

Our social, cultural and political reality is not created by powerful entities or people *out there* but by individuals, participating collectively. The future is being written and rewritten in every given moment by us. Just like our own personal lives, our collective life will be a journey of trial and error, testing different ideas, measuring their impacts and ultimately learning.

This book is not going to be an exercise in sugar-coating reality or wishful thinking. I'm not going to tell you not to worry, everything will work out just fine. We are facing a challenge of unprecedented scale and complexity. The stakes have never been higher. Our optimism needs to be humble and it demands action. It can manifest in myriad paths and alternative ways of being, seeing and thinking.

Optimism is believing that '*goodness*' – however you define it – can prevail. That the long moral arc of the universe does bend towards justice. To be optimistic fundamentally requires believing in humanity's ever-present ability to choose to live by our better natures.

~~~

Can we turn crisis into opportunity?

Some people say humanity will only come together when we meet a common enemy: consider
how Arsenal and Chelsea football fans unite when
England plays France. We have never, in human
history, needed to be so united.

Human-made climate change transcends all
boundaries, and offers us a common enemy, right
now. An eroding quality of air, soil and water will
eventually affect every single person on the planet
regardless of ethnicity, gender, religion, political dis-
position or even economic privilege. No number of
expensive bunkers can insulate people from the risk
of runaway climate change.

The data and the science* are challenging but let's
absorb that reality, because our collective concern will
be the fuel for action. We are not frogs and we are not
dinosaurs.

We are human beings living on a planet that is still
beautiful, diverse and full of life's abundant complex-
ities. It is humanity's capacity to invent, create and
care that arguably demands our saving.

~~~

A photo of my daughter licking Arctic ice, and another one of her playing with her cousin Haru, are

* Science data is in the appendix.

propped on my windowsill, in the small room where I write.

What will the planet look like in 2050, when they are young adults? Wylde and her girlfriends are already carrying the next generation in their little bodies.

The people who will be alive in 2100, perhaps even 2160, are already biologically present among us. Their eggs are playing on swings. Running. Maybe flying.

What future are we creating for them?

To be alive at this critical moment in time is both a responsibility and a privilege. We will be the generation that is either celebrated, or blamed for its apathy.

We are the ancestors of the future.

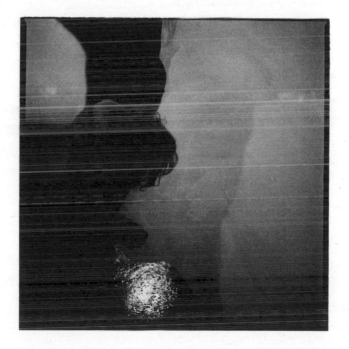

PART ONE
The Power of Money

Can we shop our way out of crisis?

The personal is political . . . you cannot corral
any aspect within your life, divorce its
implications, whether it's what you eat for
breakfast or how you say goodbye.
                    Audre Lorde, *A Burst of Light*

Too often politics is seen in a distant space. It is what
you do every four or five years when you vote in an
election, sign an online petition or, if you are really
'active', when you march or protest. These are the
explicitly political moments of our lives. But in fact,
every choice we make is intensely political. The accu-
mulation of many small actions shapes our larger
political, economic, cultural and social reality.

We cannot lament a politician's lackadaisical attitude towards climate change, and then not also
question our own lifestyle choices: how we travel,
what we eat, wear or buy. This realization can feel
overwhelming, or deeply empowering.

In capitalist economies one of the most effective
ways we can have a consistent political impact is by
examining how we speak the universal contemporary
language: money.

In the twenty-first century, money is the fuel that
powers our world, more so than gas and oil.

This section of the book will be dedicated to
the rise of what is (annoyingly) called 'conscious

consumerism', exploring the fast-moving landscape of options from food to fashion to waste, whilst acknowledging the muddy waters that these choices often lie within, and the over-riding need to simply consume less.

Numerous studies show that the majority of consumers are willing to spend more on brands if they are sustainable and protect their workers; or conversely to boycott brands that aren't. Young people are driving this change, with nine in ten millennials saying they would switch brands to one associated with a cause.

Conscious consumerism can fall into a potential elitist trap. The unfortunate, inconvenient and counter-productive truth about conscious consumerism is that the 'better', less impactful options often cost more. Sometimes way more. In fact, there's a double cost because in addition to the price, there's also the time cost of learning about the issue in the first place: perhaps by reading books like this.

We live in a false economy. The real financial costs of products and services – including the economic costs of environmental degradation – are not currently being accounted for in the price of the things we buy, but rather deferred to the future.

How much will it cost our grandchildren to clean our oceans of plastic? How much to re-plant our forests?

By not asking companies to pay for their pollution at source, we distort the present valuation of our choices. This is largely why more 'ethical' and environmentally friendly alternatives, such as organically farmed food or clothes made to last a lifetime, often appear more expensive than polluting alternatives. It means that being environmental can easily seem like an expensive lifestyle choice: the domain of those who can afford to care. (Although ironically the wealthiest people still tend to have the highest emissions.)

In Part Three we will explore potential political solutions to this paradox, looking at proposals such as carbon pricing. In the meantime, we have the option to pay a premium for 'sustainable' palm oil, hand-made, Fairtrade, or organic products, and it is important to do so, if you can afford to, to encourage the market in the right direction. With increased scale and support these alternatives are becoming more affordable for everyone.

There are other more fundamental (and ultimately cheaper) remedies: we can focus on buying less, but better. Shifting our mindset away from disposability: buying one expensive coat that lasts ten years, rather than ten cheap coats; learning to sharpen our knives rather than buying new ones; repairing what we already have; supporting vintage and second-hand; learning DIY.

Meanwhile, I am sympathetic to the argument that sustainability and capitalism are fundamentally incompatible, as capitalism has us addicted to quick product cycles that demand we buy ever more. I understand why some people claim that concepts like 'conscious consumerism' and 'sustainable fashion' are not just oxymoronic, but a distraction

Sustainable style, rather? 'Fashion fades, style is eternal,' Yves Saint Laurent once said.

from fundamentally challenging our economic and political structures.

Given the choice, I choose evolution over revolutions, which have historically often been violent and destabilizing. Whilst we still live under the umbrella of capitalism, better we realize consumers' power to reshape it. A company will live or die based on the willingness of its customers to buy its products or services, just as politicians watch and respond to market trends.

Unless you live in an *entirely* self-sufficient way, everything we touch, eat and use is connected to a complex web of trade relationships that we are responsible for being part of. When we make a purchase, we are explicitly supporting that way of doing business: by giving a company our money, we are funding them to keep going.

Of course, the power of money lies not just in how you spend it; but also, in how you earn it, save it and invest it. Many of us want to earn our money working for – or setting up – companies that are looking to solve problems. Employees often have the greatest leverage to change the working practices inside companies, and there has been a boom in social businesses, and new legal structures, that seek to remodel a new vision of capitalism.

My journey in thinking about these issues began when I was accidentally plunged head first into the epicentre of consumerism – fashion. So I'll begin there.

Chapter One
Diamonds on Your Souls

Materials, fashion and
the price of what we wear

I was always a fairly material girl: I remember trawling Portobello Market in London from a very young age, begging my mum to buy the latest trinket to catch my eye. I had a particular penchant for dressing up, and second-hand clothes were my palette. Mum's answer was usually no, met by hysteria. Sometimes my strategy worked and the market vendor would offer the object of my desire for free – sometimes my over-tired mum would give in and buy it.

Then a store opened in Kilburn selling new clothes at unbelievably cheap prices: T-shirts for £1, dresses for £3. I saved my Christmas money and would fill up one of their big black baskets until it was so heavy I could barely drag it to the cashier, let alone get it on the bus home. I never stopped to consider how clothes could be made so cheaply. It was as if they had dropped from heaven.

And one day, my life changed.

I was walking in Soho with friends when a vision of a man – blond hair, olive skin, shining as if from another world – approached us. Benjamin Hart held out a little white card and said something about modelling. My friends jumped up and down, whilst I pocketed the card and let my heart spring under a cautious face.

My mum took me into the modelling agency the first time. I was given a big blue book, which would be filled with pictures of me: I was to become my own travelling saleswoman. After we left the agency, my mum and I went to a local café and chatted in hushed and excited tones. I had just turned fourteen and had neither read a fashion magazine nor been outside Europe, so the blue book was heavy with possibilities.

Fashion offered a fairy tale for my inner child, and initially I indulged it without question: painting my face and contorting my body into strange shapes in fields, palaces and junkyards.

I was lucky enough to work with some of the world's most creative photographers, designers and artists. I made money and friends, and fulfilled my childhood dream of travelling.

Yet fashion also offered me a surprising gift: an education in capitalism.

Fur and diamonds

My questioning began with fur. There is a funny photo of me, aged around ten, in a handmade outfit, wearing a misspelt 'NO FER' sign. Then there are a few photos of me aged fifteen, before I felt bold enough to say no, walking down catwalks with the soft vestiges of a dead animal circling my neck. Eventually saying no to fur cost me work, but it was a fairly straightforward choice to make: a little box on contracts, usually appearing next to 'No Nudity'.

At seventeen, I found myself caught up in a more complex dilemma, drawn into a controversy regarding a jewellery company I was working with which had been accused of exploitation and persecution of the San – or Bushmen – in Botswana by diamond mining in their historical homeland.

To try to learn more about the situation, I met with Dr James Suzman, an anthropologist who had been living with the San on and off for over twenty years. In the midst of our intense, long conversation, James asked whether I wanted to come out to Botswana and see things for myself. My eyes lit up, and a few weeks later I was on a plane with my older sister, Elvie, headed south.

We spent ten days travelling around Botswana, discussing the impacts of diamond mining with San members, NGOs and politicians, and my mind flipped from one perspective to the next like a fish out of water.

The San had been living as gatherer-hunters on the Central Kalahari Game Reserve, the second-largest such reserve in the world, for an estimated 200,000 years. In 1997 the government of Botswana relocated three quarters of them into resettlement camps outside the reserve. Though these evictions were declared illegal in 2006, hunting had also been prohibited nationally, making the San's traditional gatherer-hunter lifestyle impossible to practise.

Successive Botswanan governments have denied it but many advocacy groups believe the reason for the San's resettlement was to enable diamond mining.

Note the inversion of the more commonplace 'hunter-gatherers', an amendment suggested by James Suzman to better reflect the relative importance of gathering over hunting in the San lifestyle (roughly 60 per cent of their diet was gathered).

Since independence in 1966, Botswana has been a politically stable multi-party democracy, and has seen decades of economic growth driven mainly by its diamond industry. It went from being one of the poorest countries in the world in the late 1960s, with a GDP per capita of $70, to having the highest GDP per capita in Africa – $18,825 – by 2015.

The situation offered a textbook <u>utilitarian dilemma</u> and forced me to consider lots of questions I couldn't answer. What are the consequences of economic development? Do the potential positives, such as investment in education and healthcare, outweigh the priceless costs of environmental and cultural degradation? Who gets to decide?

For a long time humans have connected happiness with ethics (think Hedonism or Epicureanism). But what if one person's happiness is at another's loss? Utilitarianism is the philosophy that holds that the most moral act is the one that produces the most happiness for the greatest number of people.

On that trip to Botswana, I had discovered extraordinary pieces of jewellery handmade out of ostrich eggshells. The fragility of these objects reflected the fragility of the communities behind them: it would take the San women several months to make each item, then they would typically sell for very low prices, haggled down by the odd passing traveller.

It struck me that the San's jewellery could be exported to the West and bring a good income back into the community, so James Suzman and I enlisted a friend to help set up a trading route in London. A few months later I returned to Africa to model the collection, which began to sell for much higher prices, with all the profit going back to the San. It was, accidentally, my first direct experience of Fairtrade.

This experience in Botswana marked me: compelling me to think more deeply about the impact

The photoshoot of me wearing San jewellery, by Greg Lotus, 2007.

not just of my own purchasing choices but also of my work. As the poster girl for different products, I felt an enormous responsibility to understand and respect what I was selling. I wasn't just buying a company's products, I was asking other people to buy them too.

All you need is less

Have nothing in your houses that you do not know to be useful or believe to be beautiful.
William Morris, *Hopes and Fears for Art*

Andy Warhol's infamous 1961/2 screen prints of mass-produced Campbell's Soup cans have been read both as a celebration, and as a subversive critique, of the dawn of mass-production. That same decade, the French Situationists advocated small pranks to disrupt capitalism, such as rearranging cereal boxes in a supermarket, which artist Gabriel Orozco paid heed to in his 1992 photograph *Cats and Watermelons*, balancing tins of cat food on the fruit.

Causing red lips, thin bodies, pale faces, and believed to be brought on by excessive dancing or thinking, tuberculosis was considered strangely fashionable in the nineteenth century: 'Consumption, I am aware, is a flattering malady,' wrote Charlotte Brontë in 1849, reflecting on the 'terrors of the swift messenger' that killed both of her sisters, Emily and Anne.

Say 'fashion' and most people think of clothes but fashion is symbolic of our broader relationship to the material world, and evergreen desire to own it. Consumption – once a terrible (but oddly glamorized) disease – has become a long-held trend. Go through any large city, airport or high street, and stuff and *more stuff* abounds.

A few forces have converged to paint this reality: the wage gap between 'rich' and 'poor' nations coupled with mass-production has pushed prices down; the growth logic of capitalism has demanded that companies *sell, sell, sell*; and the psychology of seasonal trends, <u>planned obsolescence</u> and increasingly sophisticated advertising has convinced most of us that we need to *buy and buy more.*

The twentieth century witnessed a cross-sector race to the bottom to manufacture cheap, disposable products and a growing appetite for them. In the US, an estimated 1 per cent of the materials used to produce our consumer goods is still in use six months after sale.

What are the consequences?

I will focus on fashion as a metaphor, but as you read through this chapter bear in mind that all consumer products have similar backstories.

Fashion can be beautiful, creative, political and fun – and we all have to wear clothes, especially when it's cold. Sometimes fashion supply chains can provide positive work opportunities that lift communities out of poverty; in other instances, they can be exploitative and even dangerous to the people they rely on. Environmentally, fashion is responsible for 10 per cent of the global carbon footprint, and 20 per cent of waste water. Three quarters of the 80–100 billion garments made every year will end up in landfill.

The fashion industry is waking up to its impact and evolving into a fairer, cleaner version of itself. In the fifteen years since I started looking at fashion supply chains, 'sustainable fashion' has gone from being a niche, sideline movement – almost considered

One of the most dastardly cases of planned obsolescence was the so-called 'Phoebus Cartel', a global network of corporations including Philips and General Electric, which conspired to limit the lifespan of light bulbs. Between the founding of the cartel in 1925 and its dissolution following the breakout of the Second World War, the average lifespan of a light bulb dropped from 2,500 hours to just 1,000.

A particularly deadly day for shopping, landing just after Thanksgiving, synonymous with sales. The term was coined by police because of the mayhem, increased traffic accidents and sometimes violence that resulted from the frenetic shopping, with one New York paper calling it 'Black-eye Friday'. Meanwhile, Jamaican musician Chronixx gives pause for thought for using the colour in a pejorative way: 'black magic, black witches … black market, black Friday … they never told us that black is beautiful.'

anti-fashion — to something essential, respected and, fundamentally, in fashion.

Yet to be brutally honest the most sustainable thing you can do is to *consume less*; mend what you own and buy second-hand. When asked what advice she would give to people who want to improve their fashion impact, sustainability expert Livia Firth's response is simple: 'Just buy less.'

In 2011, environmental clothing company Patagonia ran an infamous advert on Black Friday that read 'Do Not Buy This Jacket', with advice instead on how to reduce, reuse, repair and recycle. The advert, ironically, increased their sales. Patagonia is a cheerleader of responsible business, a certified B Corp which offers lifetime repairs on products, and gives 10 per cent of the profits to environmental charities.

There is now an abundance of online platforms selling second-hand quality clothes, circular business models with repair services, and clothes-rental companies. Indeed, the most stylish woman I know only wears vintage.

'Beyond the sustainability aspect, I find shopping vintage fun,' Giada tells me.

Then her eyes light up: 'I like finding the treasure!'

Part of the wider circular economy: keeping materials in use as long as possible, eliminating waste.

The power of no

On 23 April 2013, workers in the Rana Plaza building in Bangladesh noticed cracks in the structure. Most managers told their employees to stay at home the following day but the garment factories ordered theirs to return to work, threatening them with the loss of a month's pay if they failed to turn up. On 24 April, at least 1,135 people were crushed to death under the building when it collapsed.

One of the countless tragic aspects of the Rana Plaza disaster was that it landed, for many, as a shocking but unsurprising twist in an <u>all-too-familiar</u> <u>narrative</u>. The abuse of garment workers – aka sweatshop labour – is arguably the best-documented aspect of the perils of production.

What is the best response to this kind of news? Simply saying no?

Consumer boycotts work by affecting a company's bottom line, and also its reputation. A 2011 study found that for every day a company's boycott was in the news, its stock price declined.

Historically, boycotts have been used to push for political change. They were a central part of Mahatma Gandhi's peaceful strategy of 'non-cooperation' in protest against British colonial rule. Gandhi famously hand-spun his own clothes, and encouraged others to do so too, rather than buy British-made textiles. Similarly, during the civil rights movement, Martin Luther King called on the mass boycott of banks and shops that encouraged segregation.

In the late 1980s, an international campaign drew attention to sweatshop conditions in Indonesian factories that paid lower-than-average wages, to predominantly female workers, to make products for Nike, Reebok and Adidas. There was a surge in

In the run-up to the Rana Plaza tragedy, between 2006 and 2012, more than 500 Bangladeshi garment workers died in factory fires.

media attention on sweatshop activities and child labour throughout the 1990s, with articles increasing three- or fourfold in major world newspapers, and the implicated firms pressurized through consumer boycotts.

In response, the Indonesian government raised the minimum wage, and clothing and footwear firms signed codes of conduct committing to improve their working conditions. Looking at data from such factories in Indonesia throughout the 1990s, a team from the World Bank found that as a consequence, they distributed their profits more widely amongst their workers, and garment-factory wages nearly doubled.

Meanwhile, it must also be remembered that the garment industry has been a positive force in lifting communities out of poverty and empowering women in particular. In a study of ten countries, it was found that garment-industry wages exceeded average incomes in eight out of ten of the countries, and in half the countries resulted in three times the national average.

A study by the University of Washington which looked at different villages in Bangladesh found that in those close to garment factories, girls were nearly a third less likely to get pregnant or be married young, and more than a third more likely to go to school.

Thus, some people argue it is better to campaign companies to improve their workers' conditions, rather than to boycott the brands and thereby threaten the livelihood of their workers. Social media is empowering consumers to ask brands hard questions: Fashion Revolution's call to ask brands #whomademyclothes has been seen by hundreds of millions of people.

I always resist 'lesser of two evil' claims (i.e. that sweatshops are better than subsistence farming, or fracking is better than coal), and personally try to avoid buying into brands with negligent practices, but I think it is important to understand the nuance of the debate.

For Livia Firth, the push for fairer fashion is ultimately feminist. 'Let's establish the connection between producers and us, and be true <u>feminists</u> all the way,' she says, 'by not disregarding the millions of women abused in supply chains worldwide.'

Approximately 80 per cent of garment workers around the world are women.

Ignorance is bliss, knowledge is power (and power is responsibility)

The desire to bring to life the usually invisible people behind products and celebrate craftsmanship drove my first attempt at building a social business in 2009. Inspired by our own grandmas' generous habit of knitting for us, a friend and I founded a company called The North Circular to enlist the elderly to make hats, scarves and other knitwear.

We used eco-friendly dyes, wool from rescued sheep, and tried to make everything in the UK, using textile mills and dye sites that were legacies of Britain's once strong manufacturing industry. We put the individual knitter's name and a line of advice on each bobble hat, scarf or beanie, such as, 'Doreen says, "Don't kiss and tell."'

In 2015 we shut The North Circular down, as it had proven too difficult to compete with other knitwear companies who hand-made their products in countries where wages are lower, or simply used factories. In the aftermath, it was hard not to fundamentally question the economics of local production.

Does the exploitation of the wage gap between 'rich' and 'poor' countries drive artificially low prices, and therefore high rates of consumption for those who can afford it? Is this fair, or colonialism by another name? Does it make us any happier, and what is the environmental fallout? If consumers knew the real people behind products, would they value them differently?

'Why is everything "Made in China"?' I remember asking my mum as a child, turning over some toy or teacup at home. When people point their environmental fingers at China, angry about the exponential greenhouse gas emissions, they often forget that a large part of China's emissions are driven by the manufacture of products for export. Yet, in order for us to know what 'Made in China', or 'Italy', or 'Bangladesh', really implies, we need greater transparency.

'Opaque supply chains are devastating environments and livelihoods around the world,' says Jessi Baker, the dark-sparkly-eyed, super-smart, tech force behind Provenance, who are using technology to map supply chains. Yet she is optimistic: 'Thanks to the wealth of information now accessible, and the unprecedented consumer influence social media has enabled, our ability to unearth truths and hold companies to account has never been stronger.'

In 2014, I partnered with Provenance to run an online and physical shop that experimented with ways to tell product stories. The project raised lots of fundamental questions. How do you prioritize different ethics: animal rights, environmental, labour standards? Is the customer interested in data or stories or images?

Even for consumers who care, it is still not entirely obvious what is legitimate and what is green-washing, but labels and claims like organic, Fairtrade, carbon neutral and biodegradable help steer the way.

There is now a long list of sustainable fashion companies working hard to reconcile beauty and ethics, from EDUN which empowers artisans in Africa, to Veja and Bottletop in Brazil. Christopher Raeburn upcycles parachutes and old maps into clothes; Katharine Hamnett is the pioneer of political, organic T-shirts (and infamously met Margaret Thatcher in one); Izzy Lane rescues sheep in Yorkshire; Stella McCartney focuses on animal rights; Ninety Percent gives 90 per cent of profits back into the supply chain; Duran Lantink upcycles and made Janelle Monáe's extraordinary vagina pants . . . There are too many more to name.

Meanwhile, I also respect all the brands which don't advertise themselves as overtly 'sustainable' but emphasize craftsmanship and quality, or offer repairs. There is also a host of larger companies which have made far-reaching commitments to reduce their impact over time. In 2018, the UN launched an ambitious Fashion Industry Charter for Climate Action, with industry-wide buy-in, committing to a pathway to be <u>carbon neutral</u> by 2050. The Ellen MacArthur Foundation has made a proposal to make circular design 'the new normal': asking 20 million designers to think about how to design in a circular way.

When I dragged that heavy-black-bag-full-of-more-clothes-than-I-needed across the shop floor as a child,

Carbon neutrality can be achieved for anything from businesses and products to concerts and book tours, either by eliminating carbon emissions altogether, or by balancing them with carbon off-setting (think tree-planting) to reach net zero emissions.

I was walking with the heady dew of blissful ignorance. For a long time, we all lived in bubbles: able – and encouraged – to disconnect from the realities behind the things we were wearing, using, throwing away.

Yet the dawning age of information, driven by digital technology, is regularly bursting those bubbles; and offering us knowledge and power instead. Since we founded The North Circular, and ran the online shop, the possibilities for transparency in supply chains have exploded. In 2019, Google announced a 'moonshot for the [fashion] industry' to apply data analytics and machine learning to study raw materials, such as cotton and viscose, where 'brands have little to no visibility'. They partnered with Stella McCartney initially, with a view to creating an open source tool for the whole industry.

With technology, greater transparency is inevitable, and with that comes the increasing power for us to say no to stories that haunt, and yes to those that matter.

The power of yes

> For me, campaigning and good business is also about putting forward solutions, not just opposing destructive practices or human rights abuses.
>
> Anita Roddick

By the time I was twenty, I had pivoted my modelling work to try and promote companies I felt had an ethical backbone. This had seemed like a personally risky move to make, but I felt rewarded when I was asked to become the first global ambassador for The Body Shop: pioneers in the Fairtrade movement, or 'community trade' as they call it.

Set up by Dame Anita Roddick in 1976, The Body Shop had always insisted on not testing its products on animals (in spite of this costing them all trade in China, although PETA is 'cautiously optimistic' this law in China may soon change), emphasized the sourcing of natural ingredients, encouraged the reuse of packaging, and campaigned on social issues.

In 2012, I travelled with The Body Shop to a small village in Northern Ghana, Mbanayili, to meet the community they had been buying shea butter from since 1987 when Anita first visited. In the red clay landscape, women adorned in bright fabrics that seemed to contain the sunshine, balanced bowls of shea nuts on their heads.

The shea tree, originally a curse, now offers a blessing to the women of this polygamous community. As their property, the women were historically tasked with the hard work of grinding, roasting, kneading and clarifying the butter from the tree's nuts. Now, the global appetite for shea butter (in cosmetics) has made the women here the principal breadwinners.

In addition to giving women a voice in their society, The Body Shop dedicated a 21 per cent annual premium to community development projects. As I explored Mbanayili, the women sang as they made shea butter; a daughter ran the family shop; and the men gathered to rethatch the roofs of mud huts in rotation: taking turns to help one another, like old-fashioned <u>barn-raising</u>.

One day, intensely curious, I played journalist, camera at the ready. A woman in her seventies led me to her mud-hut home, and my girly questions built surprise bridges across our cultural gulf. Did you love your husband when you met him? Do you still? What are your dreams? One of her husband's other three wives came and sat beside her. She still has 'the love' for him, she said, laughing. Then I invited their questions. Was I married? No. Did I have children? None. How sad, they said. Suddenly the table had been turned, and I felt an obscure loneliness that contrasted with their intricately webbed community.

Before leaving Mbanayili we visited the chief's mud-hut palace for our formal goodbye. Smiling, he gave us two guinea fowl and eight yams as gifts. Trade was a gift exchange of sorts: shea butter in exchange for wells and schools. 'Long may you continue to support us, and we support you,' Madame Fati Paul, a local leader, said.

Back in the day, barns were common on farms, but roofs were heavy, so communities banded together to help raise one another's. Apparently, it was – and occasionally still is – quite a jolly and social affair. The Amish call it a 'frolic'.

It all begins raw

For centuries, people have excavated all corners of the globe, looking for the most beautiful and iridescent materials to adorn themselves and the wealthy: whalebone corsets, seal-fur coats, silk shoes and even earrings made from the heads of hummingbirds.

After a twentieth-century detour into the world of petroleum – nylon, PVC, polyester and petroleum-based fertilized crops for textiles – twenty-first-century fashion looks set to re-establish its historic relationship with the natural world, albeit on better terms.

I was travelling east on the London tube, and found myself made teary reading a report on cotton, of all things. It was by the Environmental Justice Foundation (EJF), advocating organic cotton. The benefits are widespread. Organic cotton produces up to 94 per cent less greenhouse gas emissions than conventional non-organic cotton, and requires 91 per cent less irrigation, the majority being rain fed.

This is essential considering that cotton is a very 'thirsty' crop: it can take more than 20,000 litres of water to produce a pair of jeans, and in Uzbekistan, negligent irrigation policies have almost drained the Aral Sea. As we will see in the next chapter, organic farming also offers a huge environmental solution by making the soil a <u>magic carbon sink</u>.

Our humble soil contains more than three times the carbon there is in the atmosphere, four times the amount stored in all living animals and plants and absorbs a quarter of annual fossil-fuel emissions. Soil can contain much more or less carbon, depending upon how we treat it.

The switch to organic cotton farming would also have tremendous health benefits for the estimated 77 million cotton farmers poisoned by carcinogenic pesticides every year. As I read through the case studies of some of the farmers and children affected, my mind raced through all the things I owned made of cotton – my sheets, towels, socks, shirts.

After becoming a patron of EJF, I visited India to see how an 'Earth Positive' carbon neutral T-shirt is made. We visited organic farms, wind-powered factories, and I even drank the water at a dyeing plant, where it is recycled to make sure no chemicals leak into the nearby environment. The trip made me wildly optimistic: it is possible to mass-manufacture clothes at the scale demanded by our ever-growing population, without having a negative impact.

The growing appetite for more sustainable fashion and homeware is driving an exponential growth in organic cotton: since 2004 it has doubled annually. Alongside all the early adopters, many industry giants have committed to deadlines for using fully sustainable cotton suppliers.

In 2018, I visited the world's largest sustainable fashion conference in Copenhagen, and discovered a hive of possibility. Next to the biodegradable glitter stand, you could finger through hundreds of sheets of environmentally friendly raw materials, like choosing wallpaper: regenerated cashmere, sustainable viscose, recycled ocean plastics, leathers made from mushrooms, pineapples and orange peel, and even DNA itself.

Pick your poison

I want to jump for joy that the industry is finally starting to wake up.

Stella McCartney

When sustainable fashion evangelist Stella McCartney told her parents, Paul and Linda, that she was

With roughly 100 times the environmental impact of wool, cashmere is soft to the touch but hard on the planet. Which is why labels such as Stella McCartney have started using cashmere repurposed from offcuts from Italian factories.

Made from wood pulp and prized for its soft drapey qualities that echo silk, viscose is credited as the earliest man-made fibre. It is also environmentally damaging, with over 70 million trees logged each year, often unsustainably, to produce it. Thankfully, sustainable alternatives made from properly sourced wood or alternative raw materials are increasingly available.

creating her own fashion line backed by Gucci, their first reaction was concern, having brought her up a vegetarian: 'We kind of privately thought, oh my God, what's she going to do? Because Gucci is leather. We didn't comment but we thought, that's going to be difficult,' Paul told me. Yet, he says, Stella was very clear on her stance from the beginning, telling Gucci, 'If you want me, I won't be doing leather or fur or suede.'

Stella became famous for her commitment to not using animal skins, although originally, and perhaps ironically, this meant that her label – like many others – tended towards plastics and synthetic materials that aren't great for the environment. Where possible, her company now sources recycled plastics (like Econyl, which recycles nylon fishing nets) and bio-based plastics (from wood and citrus), and they have committed to stop using virgin nylon by 2020.

The majority of fashion fibres are now derived from fossil fuels.

Stella McCartney maintains that synthetic materials, like plastics, are twenty times less environmentally impactful compared to animal leather, although she acknowledges, 'We never call our faux leathers or faux furs sustainable – we are fully aware that they derive from non-renewable resources.'

'It's a case of pick your poison,' material scientist Matt Scullin quips, comparing the ethics of leather or plastic pleather. I am downstairs in a residential basement in central San Francisco, exploring the underground laboratory MycoWorks, where a team are fermenting a form of fungus called reishi mycelium to make sheets of beautiful and durable leather from its roots.

Stella is an investor in MycoWorks's competitor, Bolt Threads, which has also developed a mushroom leather, called Mylo, and used proteins found in spiderweb to produce artificial silks without the need for boiling silkworms alive.

'Technology presents such a huge opportunity for new sustainable and circular materials to be developed,' Stella tells me. 'This is what excites me and gets me up in the morning! For me that is the future of the planet – not just in fashion, but in food and medicine.'

> In the Stone Age, we mastered natural
> materials like leather, silk and wool; the
> Plastics Age brought us synthetic polymers,
> and the Information Age unlocked life
> itself: DNA.
> Modern Meadow, www.modernmeadow.
> com/our-technology/

Modern Meadow grow leather made from cells reproduced in a Petri dish, rather than pulled from an animal's back. Beyond the ethical and environmental considerations, Modern Meadow say there are many advantages to their alternative leather, Zoa, that will make their product more appealing – such as not being inconveniently shaped like an animal.

People are also growing diamonds! Just outside San Francisco I visited the 'world's most sustainable diamond mine': a labyrinth of silver pipes and machines, baking gems. Diamond Foundry take gas from the natural gas industry, and use 100 per cent renewable energy to turn the gas – at high temperatures – into diamonds: saving the 200 to 250 tons of earth removed, the 2,011 ounces of air pollution and 143 pounds of carbon dioxide usually emitted, for a single carat of mined diamond. Diamond Foundry is the world's first diamond producer to be certified carbon neutral.

The politics of fashion

fashion **n. 1** a currently popular style of clothing, behaviour, etc.
Concise Oxford English Dictionary

My journey in fashion has taken me full circle: from the diamond mines in Botswana, to the diamond factories in San Francisco. Yet fashion is not just about where the jewels come from. Fashion is also about our behaviour and beliefs: the very act of wearing a diamond is political.

For centuries fashion has been used to display status and gender, and entrench class differences. How we present ourselves is both personal, and a cultural conversation that reflects our time, place and values: from African tribes who used jewellery to stretch their earlobes, to Victorian women who wore long dresses to hide their ankles.

Conversely, clothes have also been used to dispel hierarchies and question the status quo. Joan of Arc was burnt at the stake in 1431 for persistently wearing men's clothes. A 1620 pamphlet illustration called *The Man-Woman* satirized the trend for women to cut their hair short or dress in a masculine style, which both <u>King James I</u> and the clergy were actively trying to discourage. As late as 1919 Puerto Rican labour leader Luisa Capetillo was jailed for wearing trousers.

James I of England's outspokenness on women's hairstyles and dress sense alludes to his rather more sinister role in the popularization of witch trials in the sixteenth and seventeenth centuries. During this period many women were tortured and killed on suspicion of witchcraft, often due to accusations based on little more than their physical appearance. One particularly dangerous trait was having red hair.

HIC MVLIER:
OR,
The Man-Woman:
Being a Medicine to cure the Coltifh Difeafe of the Staggers in the *Mafculine-Feminines* of our Times.

Expreft in a briefe Declamation.

Non omnes po fumus omnes.

Miftris, will you be trim'd or truft'd?

London printed for I.T. and are to be fold at Chrift Church gate. 1620.

We express our beliefs and world view through the clothes we choose to wear: from men in suits to skate-kids, punk rockers to princesses. When Patricia

Okoumou walked into court in 2018 to be charged for climbing the Statue of Liberty as a protest against the US 'zero tolerance' immigration policies, she wore an olive dress, inscribed with the handwritten phrase, 'I really care, why won't u?'

Fashion reflects, better than most, the way our patterns of consumption are often driven by motivations that run much deeper than utilitarian need: desire, status, beauty, creativity, wealth and identity politics. What we desire, and how we want to appear to the world, also ebbs and flows like the seasonal nature of trends.

Yet the reality is that the messages and stories which our clothes tell go much deeper than the colour choices or patterns they deploy: the politics also lie in how they have been made, what impact they had on the world, and our own relationship to them. Stories that are hidden; or stories you are proud to tell.

I began my journey studying supply chains in fashion, because that was the industry that fate had landed me in, but I have always been hyper-aware that it is simply a metaphor for everything else we use, touch and buy; and our relationship to consumption more broadly.

The fashion industry has a long way to go in cleaning up its act, minimizing its environmental impact, and improving its labour standards globally, yet I am optimistic that consumer trends – and environmental necessity – are already pushing the industry in this direction. Cutting down forests or poisoning farmers is unfashionable.

We can also change our attitude to the material world: buying only what we really need or truly love; celebrating the stories behind the things we buy; and knowing the great power we wield with our yes and no.

Chapter Two
A Planetary Diet

Meat, farming and
the food of the future

I come from a farming family. My dad was a fisherman, and my grandparents' farm in south Wales had been producing free-range eggs long before that concept even existed. They had a horse, and my mum recalls that her father would 'take the eggs – slung over the horse in big trays – down the mountain to the valley'.

When I was ten, I finally computed that you have to kill an animal to eat it, and became one of those annoying, package-reading, proselytizing vegetarians.

'How do you get your protein?' was the mantra I'd endlessly meet.

'Chicken?' was the response in Africa.

'I added some ham for free!' in France.

Being vegetarian usually felt like an awkward, and fundamentally unhealthy, choice to make.

Yet, over time, the question has changed.

'Are you doing it for health reasons?'

'Or for the environment?'

Better still is the silence.

Chefs no longer tend to throw ham into my soup as a gift in continental Europe.

The new challenge is trying to get them not to add the butter.

I'll eat eggs from chickens I know – if they are in
someone's backyard and they are wandering
around and they're happy and they don't have
a husband.

Ellen DeGeneres, *Relatable* (television programme)

Like Ellen, I've been some version of vegetarian,
pescatarian, or vaguely vegan (vagan/pagan anyone?)
for over twenty years: oscillating as I have debated the
implications of dairy farming, the air miles of tropical
vegetables, sustainable fishing, or whether free-range
is free enough.

Unlike Ellen, I'm not a comedian and am aware
that talking about veganism (even vaganism) is distinctly unfunny – polarizing, even irritating. Yet as
animal agriculture is increasingly understood to be at
the centre of our climate crisis, winning first or second place – and offers the simplest, lowest-hanging
fruit for individuals to act on – it feels essential to
explore. Bear with me.

Depending on how they are calculated, estimates for animal agriculture's global greenhouse gas emissions have ranged from 18 per cent (UN FAO, 2006) to 51 per cent (Worldwatch, 2009).

The art of meat

Grub first, then ethics.

Bertolt Brecht

We owe a lot to meat, including the birth of culture.
According to (my now friend) James Suzman, humans
were once much like gorillas, spending most of our
time looking for food to graze on, then digesting it.

Then, with our mastery of fire, came an energy break-through: our ability to cook and eat small chunks of meat. Eating cooked meat enabled our brains to grow, and also gave us free time to evolve what are now quintessentially human traits: to contemplate, make music, and draw on walls.

I could get on board with the way gatherer-hunter communities killed and consumed meat. Were I a gatherer-hunter, I'd almost certainly be eating it, and when I spent time with the San, I did (much to the peril of my digestive tract). Yet we are a long way from the communities that killed rarely, ate or used every part of the animal and wasted nothing. Today, through indus-trial farming, we kill a billion animals a week to eat.

A second significant breakthrough in human evolution is accredited to the 'agricultural revolution' when nomadic tribes settled, and began to domesticate animals and plants. This allowed us to produce larger quantities – although a smaller range – of food. 'Wheat, rice and potatoes domesticated *Homo sapiens*, rather than vice versa,' quips historian Yuval Noah Harari.

The impacts of the Agricultural Revolution were multifaceted, and it has been credited with everything from human population growth, to the formation of private property, centralized political structures, hier-archical ideologies, and 'civilization' as we know it today.

The Agricultural Revolution also fundamentally changed the planet's population. Ten thousand years ago, humans made up 1 per cent of the weight of verteb-rate land animals: the rest were all wild. Today, wild ani-mals make up just 1 per cent – the other 99 per cent is

A species, much like other animal species, but with complex language skills, the ability to plan into the future, and a commonly held interest in fancy objects and self-decoration. I highly recom-mend Yuval Noah Harari's book *Sapiens* if you want to understand this species better.

accounted for by humans, our farmed animals and our pets. We have killed, caged and domesticated so many creatures that now wild animals are fetishized: an increasingly rare and exotic experience to visit on safaris.

As the most fundamental of human requirements, food and water represent the natural limits to population and economic growth: agriculture already uses nearly half of the world's ice-free land, and 70 per cent of our fresh-water supplies.

Food production represents one of the biggest sore spots in the nexus of interconnecting issues which we face – deforestation, habitat and species loss, climate change, pollution of land and water, heavy antibiotic use, and public health – making it a promising candidate for scalable impact.

And there is a quiet revolution under way.

Responding to technological advances and environmental risks, scientists and entrepreneurs are rethinking how we can optimize the growth of food, and create alternative meats, to use dramatically less land and water – and potentially improve animal welfare and nutrition along the way. We are now entering a new agricultural revolution that might see farms become vertical, and animals-as-meat obsolete.

Beyond meat

If slaughterhouses had glass walls,
we'd all be vegetarian.
Linda McCartney, *Linda's Kitchen*

Linda McCartney's pioneering vegetarian burgers and sausages have an iconic place in my childhood memory, as powerful as the music associated with the name McCartney. A few years ago, I went to meet Linda's partner, Paul, at his office in London, and he told me with bright eyes how that journey unfolded.

'The whole idea of reprogramming yourself is really exciting,' he said. Linda and Paul were in their thirties, living on their Scottish farm, when they decided to try being vegetarian. They had been watching lambs playing in the fields, whilst tucking into a plate of lamb, when they made the connection.

'We are on this little sphere in space, between Mars, Venus and the perfect distance from the sun,' reflected Paul. 'We find ourselves amazingly positioned with these fellow creatures in time and space – let's give them a break.'

At first the idea of being vegetarian felt daunting: 'I remember going out with Linda's dad to a posh dinner to Claridge's. We said, well, we're vegetarian, and the waiter looked down his nose, and said, we'll try to think of something: well, it was just a plate of vegetables. We went to France and they did not even know what you were talking about.'

Being vegetarian seemed to suggest 'a hole in the middle of their plates' and it certainly wasn't a mainstream idea, but they immediately found it fun and 'started gradually just to fill that hole with ideas'. One such idea was an alternative burger made from vegetable protein, which they tested at home. This eventually led to Linda forming her own alternative-meat food company – Linda McCartney Foods – in 1991. 'The whole idea just started to catch on,' reflects Paul on its success.

The McCartney family were pioneers in what is now a worldwide trend: the global substitute-meat market is a multi-billion-dollar industry growing at around 8 to 10 per cent annually, and pivoting in surprising and sci-fi ways. This is being driven by the groundswell towards vegan, vegetarian or '<u>flexitarian</u>' diets. Veganism has grown over 500 per cent in the US: and according to the US Department of Agriculture, since 2004 meat consumption has been steadily declining.

This portmanteau denotes a largely vegetarian diet with occasional meat dishes. In essence, a more environmentally conscious approach to being an omnivore.

Conscious consumerism, which arguably began in food, changes the narrative of capitalism: from the story of humanity as innately selfish and greedy beings, to one in which people will pay extra, or make perceived 'sacrifices' because they care. Yet for every well-intentioned person who opts to eat less meat, there is also a growing global population and rising middle class, increasing the demand.

Pat Brown, the founder of Impossible Foods, certainly doesn't hedge his bets for reducing meat's impact on the planet by relying on conscious consumers. 'We are not interested in vegans and vegetarians,' he told me candidly. Instead he thinks the only way to turn this crisis around is to divorce meat from animals: most people 'love meat *in spite* of the fact that it's made from animals, not because of it'.

I first met Pat in our office in London in 2016 after it emerged that we shared a company name (more on that later) and we were amicably negotiating trademark boundaries for 'Impossible'. He was a bio-medical researcher at Stanford when he landed on the idea for Impossible Foods, because he was looking for 'the most important problem in the world, that I can do something to help fix', and decided the answers lay in the meat industry. 'I realized that, by a whopping margin, the catastrophic environmental impact of animals as a food technology was like number 1, 2, 3, 4 and 5 on the list.'

Whilst the aeroplane has become the poster child symbol of environmentally destructive life-style choices, animal agriculture generates more greenhouse gases than all forms of transportation combined. It is the leading source of methane, and nitrous oxide emissions, which are respectively 34 times and 310 times more powerful than CO_2 as a greenhouse gas.

Livestock requires disproportionately large areas of fresh water and land – 83 per cent of all agricultural land – which destroys habitats and forests, reducing biodiversity, and is the primary driver of deforestation around the world (91 per cent in the Amazon).

Is it the leading cause of the climate crisis? The FAO attest that livestock emit 14.5 per cent of green-house emissions, second only to fossil fuels, whilst the Worldwatch Institute make the case that it is ac-tually the number one contributor: creating 'at least' 51 per cent of emissions. Whether animal agriculture wins first or second place, it is clear the impact is enormous.

In 2017, over 15,000 world scientists – including many Nobel Laureates – signed a 'Warning to Humanity' calling for us to reduce our per-capita consumption of meat in an effort to curb environmental degradation.

The following year, scientists from Oxford University behind the most comprehensive study of the impacts of farming on the environment concluded that the 'single biggest' way to reduce environmental impact on the planet is to avoid eating meat and dairy products.

As the Impossible Foods team took over our office kitchen and cooked up a selection of mini burgers, Pat told us how he, and his team of scientists, had separated out the different molecules in meat and found equivalents in plant ingredients: 'The vast majority of all animals on Earth get all their protein from leaves. There is plenty of protein there – you just have to eat a lot of leaves.'

The result is a vegan burger that 'bleeds'. Numerous blind taste tests suggest it is indistinguishable from meat, but it uses ten times less fertilizer and water, over twenty times less land, and emits ten times less greenhouse gas than the average American beefburger.

Three years later, I found myself at Impossible Foods' research lab, tasting burger 2.0. We wandered past large aquarium-like windows as scientists in hairnets and white coats pumped 'protein gels' into trays.

Impossible Foods claim their burger is healthier than the factory-farmed beefburgers that make up 99 per cent of the beef eaten in the US: it contains as much protein and iron, 18 per cent fewer calories, less saturated fat, no cholesterol, no antibiotics and no hormones. This is notable considering that the pervasiveness of antibiotic use in conventional farming is now understood to be affecting bacteria's resilience to them, creating 'superbugs' in an epidemic that by 2050 could threaten humans in greater numbers than cancer.

Pat was wearing a well-loved, faded T-shirt with a drawing of a cow with a red line through it, like a no-smoking sign. In his hand was a reusable water bottle with an 'I love GMO' sticker attached. Genetically modified foods have in the past been dubbed 'Frankenstein foods' by the media and treated with enormous suspicion by consumers but as Pat points out, we have been genetically manipulating plants and animals for centuries, and most cheese is produced using genetically engineered yeast, as is insulin for diabetics.

For much the same reason that nobody bothers to specify what they mean when they say W W W or BBC or UFO, the initialism for Genetically modified organisms is so embedded in the popular psyche that it hardly needs explaining at all. Although please note that, depending on who you are talking to, GMO could be mistaken for the Glenn Miller Orchestra, a popular swing-dance band from the 1930s.

Impossible Foods' strategy is largely economic. Currently, their burgers are priced competitively with premium cow meat, but their aim is to ultimately compete with commodity beef, and in 2019 they began selling in Burger King. 'When we have a product which is delicious, better nutrition, and is *cheaper* than the cow version, that's when the wheels come off the beef industry. If we can take a low double-digit fraction of the market for ground beef, the beef industry will be at the tipping point.'

Pat compares the meat and dairy industry to the 'horse and carriage' or film technology, before they were disrupted by cars and digital cameras. Impossible Foods are 'going after beef first' and then plan to work on all other types of animal products: likely prioritizing fish next, because 'there is a catastrophic meltdown in fish populations'. By working with other entrepreneurs, and potentially making their technology open source in future, they aim to 'completely replace animals in the food system by 2035'.

Lab-grown meat

> We shall escape the absurdity of growing a
> whole chicken in order to eat the breast or
> wing, by growing these parts separately
> under a suitable medium.
> Winston Churchill, *Strand Magazine*,
> 1931, 'Fifty years hence'

In announcing this seemingly audacious goal, Pat knows he is not alone in his mission. There are many other companies, such as Beyond Meat, competing to produce plant-based alternative-animal products: using everything from pea proteins, algae, mung beans and beetroot juice to produce fake meats, vegan mayonnaises and eggs.

More controversially, scientists around the world have been experimenting with growing cultured animal products in laboratories. The process requires extracting cells from an animal and then feeding those cells a nutrient-rich broth in a laboratory so that they can multiply into a product: similar to how Modern Meadow are growing leather.

In 2003, Oron Catts and Ionat Zurr produced the first lab-grown meat from frog cells. A decade later, the world's first lab-grown burger was developed by Professor Mark Post and a team of scientists at Maastricht University in the Netherlands. They took stem cells from a cow and grew them into strips of muscle, which they combined and cooked to make a burger. Thirty laboratories around the world and multiple companies are working on developing cultured meats, and it is expected they will reach the market in the next few years: Memphis Meats and Finless Foods, to name two.

A lot of biotechnology (including 'cellular agriculture', 'cultured meat', 'labriculture', or 'clean meat' as it is sometimes called) has been met by fear and resistance from the public, but the CEO of the Biotechnology Industry Organization, James Greenwood, says its implications for healthcare and the environment will be profound. Biotechnology is even being used to try and bring back lost species, such as the *Rheobatrachus* frog which has been extinct since 1983, and the Tasmanian tiger.

'This is the most transformational human endeavour in history,' Greenwood says. 'We are a major part of the answer to questions about mankind's ability to live sustainably on this planet.' Bruce Friedrich, co-founder of the Good Food Institute, says, 'It's likely that, in a few generations, animal slaughter for food will be extremely rare in the developed world.'

Lab-grown meat proposes to solve our environmental challenges without demanding mass behaviour change: it is understood to produce 80 to 95 per cent less greenhouse gases, use 99 per cent less land and 80 to 90 per cent less water, than conventional meat. However, lab-grown meat currently predominantly uses Fetal Cow Blood (FCB), so it is still not yet divorced from industrial animal agriculture, although companies are racing to produce without FCB and some say they can. It has also come to light that the energy used to power the manufacturing process may be higher than that of some conventional meat sources.

As awful as it sounds – extracted from the foetuses of pregnant cows taken to slaughter.

Like Impossible Foods, lab-grown meat also relies on industrial, chemical-heavy, mono-crops (such as corn or soya) for 'feed' that many environmentalists argue is another essential part of our agricultural problem. When Abi Aspen began working as a lab-meat scientist, whilst writing her PhD on cellular agriculture, she was optimistic it could 'save the world', but she soon came to realize they were 'feeding really shitty grown crops to cells instead of animals'. Abi left her job, and is now focused on regenerative grains and horticulture. She was out in a field, picking cauliflower, when I called her. 'Lab-grown meat is not a solution to the problem,' she told me, 'it's just a plaster on a gaping wound. We need to think more systemically.'

Slow food

Pioneering chef and activist Alice Waters represents an alternative 'slow' food narrative, which is deeply suspicious of GMO and non-traditional farming methods.

Her approach is deeply philosophical. 'The way that you eat becomes the way that you think,' she tells me. 'When you are eating fast food, you are digesting the values that come with the food: the idea that it's okay to eat in your car; the idea that more is better; the idea that time is money; the ideas that cooking and farming are drudgery. These ideas all come from a fast food industry that wants you to forget about the seasons, wants you to believe that you can have anything, anytime, anywhere in the world. These are the ideas that are destroying the planet and our human values.'

Alice grew up during the Second World War, and her parents had a 'victory garden', producing all the food they ate. She is now 'absolutely rigid' on only eating local, organic and seasonal produce: research-ing slow food restaurants when she travels, and taking her own food on planes. 'I find eating seasonally a huge pleasure,' she says, 'being denied tomatoes all year long until they are great and then I eat them every day and I am so happy. I think if you eat second-rate fruits and vegetables all year long, when the good thing comes around you're bored – you don't even notice it.'

Alice makes exceptions to her local rule for spices, cocoa, tea and coffee, but only buys them if she knows 'they are completely sustainable Fairtrade'. When we visited Alice for Easter lunch, my daughter hid organic, Fairtrade chocolates in her garden (which she was very excited to find again a few hours later). Alice and I tried to decipher the label on the back of a carrot-shaped bar: 'This chocolate is certified 72 per cent Fair for Life,' it read!

Alice has been advocating the slow food movement in her discreet, Japanese-veranda style, wooden restaurant in Berkeley California since the 1970s, long before the idea became so popular. The menu at Chez Panisse offers lots of meat options, albeit all from local, organic farms. 'We are probably a bit too meat heavy and I want to change that,' she acknowledges, before musing on the loyalty she feels to support the farmers she has long worked with.

Magic soil

A nation that destroys its soils, destroys itself.
Franklin D. Roosevelt, February 1937,
letter to state governors on a Uniform
Soil Conservation Law

Alice's approach – and the slow food movement she aligns with – raises a fundamental question. Can animals ever be raised to be eaten in a way that is environmentally friendly, even *good* for the environment? There are many farmers who argue that livestock are important components to sustainability on farms, because their manure helps to fertilize the land, improving soil quality which in turn acts as a carbon sink: naturally absorbing carbon from the atmosphere.

Patrick Holden, founder of the Sustainable Food Trust, argues that rotated livestock can actually help environmentally: through 'mob grazing' where the animals mimic wild patterns by moving regularly, so that grasses can replenish, and naturally capture carbon.

'If we are to move from chemical industrial farming to regenerative farming, animals have to be part of the system. They have been involved in the water, soil, carbon, nutrient cycles for millions of years, they have evolved alongside vegetation, and we can't simply write them out of the equation. Their dung and urine, the way they graze and browse, and trample vegetation into the soil so it gets absorbed again, are all vital processes for soil health,' Isabella Tree, author of *Wilding*, tells me. Abi Aspen agrees animals play important roles in positive farming, although amusingly points out, 'That doesn't mean you have to eat them!'

Some farmers argue that cows only become an environmental 'problem' when factory farmed, as the methane they produce would normally be naturally absorbed by bacteria in healthy soil. Whilst other entrepreneurs and MIT scientists have found, rather surreally, that feeding cows seaweed more than halves the amount of <u>methane</u> they produce.

So, environmentally speaking, is it better to eat a locally raised, grass-fed (perhaps even sea-grass-fed) organic piece of meat, or a soya derivative that comes from another continent? A future that includes a few pasture-raised animals is certainly a more romantic image than large meat labs.

> From where I stand now this bloody carcass of
> roadkill deer in front of me feels more vegan
> than the plastic packet of cacao nibs I once sold.
> Well, if not more vegan, at least more honest.
> Mark Boyle

Yet, when I put this to Pat Brown, he quickly dismissed these claims as 'propaganda', 'a complete fraud' and (ironically) 'complete bullshit'. Pat's main concern with the environmental impact of animal agricultural industry is the high land use it requires – and grass-fed animals require much more land than factory-farmed animals: 'Every grass-fed cow is a net negative impact on the environment. Full stop.'

A report published by the University of Oxford's Food Climate Research Network concludes that 'grass-fed livestock are not a climate solution'. Although there is some locally specific evidence of carbon sequestration of grass-fed animals, that effect is apparently time limited, reversible and outweighed by other greenhouse gases generated: 'The inescapable conclusion of this report is that while grazing livestock

A greenhouse gas that's roughly thirty-four times more environmentally damaging than CO_2 (though only lasts about twelve years in the atmosphere) and near-synonymous with cows, which release around 30 to 50 gallons of the stuff a day. It is a popular misconception that the majority of methane produced by cows comes from their incessant farting when, in fact, 90 to 95 per cent actually comes from burping.

have their place in a sustainable food system, that place is limited. Ultimately, if high-consuming individuals and countries want to do something positive for the climate, maintaining their current consumption levels but simply switching to grass-fed beef is not a solution. Eating less meat, of all types, is,' the report concludes.

Pat argues that instead of comparing intensively farmed land to sustainably farmed animal agriculture, we should be considering the 'opportunity cost' of land use: i.e. how much carbon we could capture if we freed up the 83 per cent of farmland currently used for animal agriculture, to re-wild it? His point is, mob-grazing cows offer a poor substitute for the biodiversity of the wild.

Not only is agriculture responsible for the vast majority of deforestation but, without it, existing global farmland could be reduced by an area the size of China, the EU, the US and Australia combined. Freeing up this farmland offers an enormous opportunity to restore biodiversity and wildlife, allow tropical forests to regenerate, and greater sequestration of carbon in the soil. A Harvard and NYU study by Matthew Hayek calculated that re-wilding this land would sequester carbon dioxide equivalent to the last sixteen to nineteen years of global fossil fuel emissions.

In his 2016 book of the same name, legendary biologist E. O. Wilson set out the argument that half of the Earth's land should be given over to a human-free nature reserve in order to preserve biodiversity. Wilson specializes in myrmecology, the study of ants.

Tending the wild

Freeing up and re-wilding all that land, as per Pat's vision and E. O. Wilson's Half-Earth movement, is a simple, obvious and effective solution that offers to solve many of our multifaceted environmental problems, but can re-wilding be economically and politically viable? Are there ways that farmers can re-wild their land, whilst retaining their livelihoods? Can we produce food from 'wild' land? Meanwhile, how are we going to grow our grains and vegetables?

When I asked Hayek how this transition might be achieved economically, he said, 'One option may be for governments to pay low-yield producers to abandon their agricultural lands if they're willing.' Of the global agricultural subsidies – worth $1 million a minute – only 1 per cent currently benefits the environment. Most of them drive problematic ecological practices: deforestation, soil erosion, high chemical inputs. Numerous reports have argued that a large part of these subsidies could be redirected to pay farmers for ecological services instead, or help guide them towards agro-ecological farming practices. 'Small producers need to be supported,' my friend Ben, an organic farmer and forest gardener, tells me.

In 2015, France's agricultural ministry announced the '4 per 1,000' initiative: encouraging agro-ecological practices that would aim to increase the organic matter in soil, to capture more carbon. The United Kingdom's 2020 Agricultural Bill offers to be game-changing in this regard.

Meanwhile, some private individuals are taking the initiative, and rethinking how they use their own land. Ray Chan bought a 133-hectare piece of land in Alto Paraíso, Brazil, that was previously a cow pasture, and

A silver lining for every British Remainer: the post-Brexit 2020 UK Agricultural Bill seeks to align agricultural subsidies with environmental improvements, so that farmers are rewarded for creating 'public goods' such as better air and water quality (as opposed to the EU subsidies which created a perverse incentive to farm every square inch of land).

Agro-forestry, put very simply, is incorporating trees into agriculture. Put trees on your farm and hey presto, you've got healthier soils, higher crop yields and increased habitats for wildlife. It forms part of the wider practice of agro-ecology, which generally aims to work with nature, and not against, for sustainable farming practices.

developed it according to 'agro-forestry' principles. Three years later the land supports 10,000 fruit trees, vegetables, fruits, medicinal plants, and chickens for their own egg consumption.

After seventeen years trying to run their 3,500-acre inherited estate, Knepp, as a failing farm in southern England, Isabella Tree and Charlie Burrell decided to do something radical: they reintroduced herbivores that could roam freely and then let nature take over. Many years later, the infamous site of 're-wilding' feels alive, and full of birdsong. Knepp estate has become home to many rare species including turtledoves, nightingales, cuckoos, lesser spotted woodpeckers, peregrine falcons and bats: 'They've all found us on their own,' Isabella says.

The only significant management they do in this re-wilded system is to cull the large herbivores, because ordinarily they would die through predation (by, for example, wolves before we wiped them out in the UK) or starvation. 'What you want is enough animals to stop the scrub turning into closed-canopy woodland (which is species poor and very undynamic) and not so many animals that they over-graze and over-browse, destroying the scrubland and reducing the habitat to species-poor grassland,' Isabella tells me. Knepp consequently produces 75 tons a year of 'sustainable, ethical, pasture-fed organic Longhorn beef and venison and pork'.

After twenty years of research, and interviews with indigenous elders, in *Tending the Wild* Kat Anderson makes the case that contrary to the stereotypical image of gatherer-hunters roaming a vast wilderness, or of humans as inherent destroyers of biodiversity, Native Americans were active stewards of their land. Through centuries of intelligent land management – such as harvesting, pruning, sowing, and the intentional use of fire – the indigenous communities of Northern America were able to *increase* the biodiversity of the land in which they lived.

After visiting Knepp, I entered into a lengthy and generous email exchange with Isabella. Whilst she believes we have to eat many fewer animal products, she also believes that animals still have an important role to play in ensuring we grow vegetables and grains in a sustainable manner. She is also anxious about the industrial, chemical agriculture and mono-crops that sit behind many of the meat alternatives.

'The priority is to get rid of factory-farmed, grain-fed meat, and industrial chemical agriculture,' says Isabella, 'but we have to be careful about what we embrace in its stead and ensure we're not kicking off other unsustainable systems.'

So perhaps there is an environmentally sensitive vision that affords us some well-sourced animal products, but it would certainly have to be a lot less of them. Indeed, this was the vision for a 'Planetary Health Diet' proposed by scientists in the *Lancet*: they argue it is possible to feed 9.6 billion people (the expected population of 2050) in a sustainable and healthy way, if we transform our diets to eat significantly fewer animal products and less sugar, and halve food waste. The Harvard/NYU study argued

According to the UN, if food waste were a country, it would be the third-largest emitter after China and the USA. (Though it would have stiff competition with cement/concrete, which also claims that statistical throne.)

that following this diet could lead to ecosystem restoration that captures the past five to eight years of fossil fuel emissions.

Beyond chemicals

> As crude a weapon as the cave man's club,
> the chemical barrage has been hurled against
> the fabric of life.
>
> Rachel Carson, *Silent Spring*

Organic farming offers both to restore soil health and to capture carbon. Beginning in 1981, the Rodale Institute conducted the longest-running trial of its kind, to look at the environmental impact of organic farming methods, comparing organically fertilized fields with conventionally fertilized fields on its 330-acre farm in Pennsylvania.

After thirty years they published their findings: claiming that a global switch to organic farming methods could sequester more than 100 per cent of current annual emissions of carbon dioxide in the soil. 'Organic farming may be one of the most powerful tools in the fight against global warming,' they concluded.

Chemical fertilizers and pesticides must have appeared utopian in the mid-nineteenth century when they were invented. Along with industrial processes, antibiotics and monocultures, they have harvested an extraordinary boom in agricultural production, helping to feed an exploding human population.

Yet as the ecological impacts of industrial agriculture are becoming clearer – depleted soil quality, chemical pollution, ground water pollution, high carbon impact, fossil fuel dependency, dangerous declines in insect populations – the value of chemical farming inputs is being re-evaluated. Our soils are now so depleted of nutrients that the UN's Food and Agriculture Organization (FAO) estimates we may have as few as sixty harvests left. 'Is it really true?' asks psychologist Anouchka Grose, exploring the rise in 'ecological grief'.

Organic began in the 1970s in the United States as a disorganized and decentralized movement with different methods for certification, but it is no longer niche. By 2015, 84 per cent of Americans were occasionally buying organic food.

Meanwhile, innovations in agricultural technology suggest potential breakthroughs in fertilizers. Harvard chemist Dan Nocera has developed a 'living biofertilizer' which not only doesn't produce greenhouse gases (a potential saving of the 2 per cent global energy requirements of conventional fertilizers), but is actually carbon negative – absorbing more carbon from the atmosphere than it uses.

Organic farming has often been dismissed as a nice but impractical idea because, historically, it has produced smaller yields. In 1971 Earl Butz, then US Secretary of Agriculture, declared, 'Before we go back to organic agriculture in this country, somebody must decide which 50 million Americans we are going to let starve or go hungry.'

It's an essential question: can organic farming methods meet our existing and growing needs for food? The answer seems to be yes, though there's a caveat. According to a study in *Nature*, organic farming could provide enough food to feed the predicted population of 9.6 billion people in 2050 without expanding the area of farmland currently used – i.e. whilst preserving what little wildland we have left – *but only if people lower their meat consumption.*

Conversely, a twenty-two-year farming trial, studied by Cornell University, suggests that organic farming *can* offer the same yields of soya and corn, whilst reducing energy requirements by 30 per cent, locking more carbon in the ground, and critically reducing water consumption.

When depicted on the big screen, Artificial Intelligence usually has quite catastrophic consequences. See: *Blade Runner, 2001: A Space Odyssey, Ex Machina, The Matrix...*

Meanwhile, perhaps, technology can empower traditional techniques. A paper advocating the role Artificial Intelligence can play in climate solutions argues, 'Much of modern-day agriculture is dominated by monoculture . . . [but] robots run on machine-learning software could help farmers manage a mix of crops more effectively at scale, while algorithms could help farmers predict what crops to plant when, regenerating the health of their land and reducing the need for fertilizers.'

These kinds of changes might make regenerative farming and organic food more accessibly priced. Meanwhile, although organic food can be prohibitively expensive, eating less meat is often more cost effective. Foodies like Jack Monroe and Haile Thomas have advocated low-cost plant-based meal recipes, such as Monroe's '9p' vegan burger, or Thomas's non-profit HAPPY which provides low-cost vegan education to underserved communities.

Water is life

'How do we divorce the historical connection between growing population and rising incomes, and the stresses that places on land and water, because we have now completely outstripped the planet's ability to sustain the number of people that are on it, at 7.5 billion?'

That's the question that Matt Barnard was pondering when he decided to found the first indoor organic hydroponic farm in the world, Plenty, which grows leafy greens, berries, tomatoes and carrots in vertical towers.

After donning boots, gloves and lab coats, and walking through two air baths and pools of bleach, we entered a brightly lit indoor space, with towering walls alive with greenery. Barnard pulled off a selection of green leaves for me to try: some spicy, some sweet, some colourful. Did I like the taste? he asked me. Which did I prefer? Matt seemed seriously concerned to know, as trying to make greens 'craveable' is a core part of their focus.

Plenty's crops are grown in small amounts of soil in this indoor farm, then transplanted into the walls, where they are fed nutrients through the water (a hydroponic system). LED lights pretend to be the sun, and the farm is monitored and moderated by AI robots, which adjust the levels of photosynthesis, nutrients and air, to optimize the plants' yield and flavour.

Indoor farms are more energy intensive, and so will require renewable energy supplies before they can make comprehensive sustainability claims. Yet as a consequence of this highly controlled and techno-logical environment, Plenty have been able to avoid using pesticides and dramatically increase yields; they use less than 1 per cent of the land required by regular agriculture, and less than 5 per cent of the water.

Having previously been an investor in water tech-nologies, a fear of water scarcity was Matt's initial motivation for setting up Plenty. 'Water is the thing that is most essential to life. The water footprint is a big deal to us and one of the primary reasons we exist, as the water system is under such serious stress,' he says. 'I view water scarcity as an emergency that's al-ready here but we're just not acknowledging.'

Only 1 per cent of the global water supply is accessible fresh water, 92 per cent of which is used for agriculture. Already over 2 billion people lack access to safe drinking water, NASA have found that over half of the world's aquifers are being depleted at unsustainable rates, and many rivers have been so heavily utilized for agriculture, they no longer reach the sea.

Water is life. Around the world, the expanding regions of drought – exacerbated by climate change – are becoming a source of conflict and contention

Giant underground sponges that hold water, and provide most of the world's drinking and agri-cultural water. The Ogallala Aquifer lies under eight states in the Great Plains, USA, and provides a third of the country's irrigation. It risks being 70 per cent depleted by 2060, but would take thousands of years to naturally re-fill. The world will likely need more desalination plants, but they are ex-pensive and energy intensive.

amongst communities. The UN estimate that half of the world's population live in potentially water-scarce areas, and Syria's drought was attributed as a major cause of their civil war.

Thus, a technological revolution in how we grow food – to use less water and land – is welcome news. Matt's vision is to see indoor farms positioned closer to cities, reducing the thousands of miles that food often travels, and therefore extending food's shelf life, which he hopes will also reduce food waste. One day, perhaps, we will have miniature hydroponic farm units in our homes.

The Gleaners

The only thing that is going to save the planet, is falling in love with nature.

Alice Waters

'They might be thirsty!' my two-year-old daughter once screeched, trying to wake me up and hurry me down to help her water the tomatoes. We had just moved to the countryside from the city, and were spending weekends in the greenhouse. At first she found gardening 'boring', proclaiming that she didn't like mud, and would relentlessly try to persuade us to take her back inside. But as the small green buds pushed up through the earth, growing taller by the day, she began to call me excitedly over to them, to see the changes.

Like organic farming, but a bit more esoteric. Developed by the Austrian educational philosopher, architect and clairvoyant Rudolf Steiner in the 1920s, many of its practices (use of manure instead of chemical fertilizers, crop diversity, the treatment of animals, crops and soil as a unified system) were ahead of their time and are still widely used today. Others (burying a cow's horn stuffed with ground quartz to imbue the soil with cosmic forces) are disputed.

Consider watching *The Gleaners and I,* a documentary by the late Agnés Varda, for more insight into the lives of those who 'pick something up so nothing gets wasted'.

Planting seeds in the ground and watching giant pumpkins or courgettes emerge is an ordinary kind of magic, even for an adult, and somehow the food tastes better. The vegetable patch creates a colourful sculpture garden of diverse shapes, sizes, twists and curves that you don't normally find in supermarkets. Would you ever *not eat* a vegetable you had grown because it was 'imperfectly' formed?

The area we moved to has two biodynamic farms, where I can buy local, organic and thoughtfully grown food. As I finished writing this chapter, we went to an annual apple-pressing gathering nearby: apples had been 'gleaned' from local harvests (left unpicked because of their imperfections). Everyone chipped in, throwing apples into the pressing machines, to make juice (and later cider). Food has the capacity not just to feed, but to create community.

The event's organizer, Tristram Stuart, turned up with kilos of apples from a local field to add to the pile. He told me about the organization he founded, Feedback, which deploys volunteers around the country to pick up food that will otherwise be left to rot. Currently, 25 to 50 per cent of food perishes, or is thrown out for being cosmetically imperfect, before it reaches the shop floor. Meanwhile, a large part of landfill is food waste, where it produces methane.

This is not a small matter. Project Drawdown places 'reduced food waste' and 'plant-rich diets' as the third- and fourth-biggest measures that can be taken to combat emissions and climate change.

74

Different initiatives have sought to combat food waste: campaigns which celebrate (and sell) seemingly imperfectly shaped fruit and vegetables; nose-to-tail meat dining; food-sharing apps like Olio and Too Good To Go, and gleaning/foraging networks. Alice Waters is turning problem into opportunity: leading a new campaign to connect schools to farms, so that food scraps are used as compost, to support regenerative agriculture, rather than ending up in landfill. 'We already produce enough food for a growing population,' Tristram told me. 'We just have to stop wasting it.'

Regenerative agriculture is number 11 in Project Drawdown's list of climate solutions: they advocate using a mix of techniques for soil health and carbon capture – no tillage, no external nutrients, no pesticides or fertilizer and mixed crop rotations.

Embracing nuance at breakfast

So, could the biggest aspect of our environmental challenge be the easiest to fix? Is our climate crisis as simple as deciding what to eat? If only it felt simple.

We need to think about food. We might one day be able to stop burning fossil fuels, but – unless you are a breatharian – we won't stop eating. Food has the capacity to destroy our land, threatening our own survival, or indeed, to potentially be at the forefront of efforts to rebuild our natural world. What we choose to eat, if we are lucky enough to have choice, is deeply personal, subjective, and cultural. The impact of our diets presents an insanely confusing landscape, with lots of competing, often contradictory, information, polarized beliefs and hidden agendas.

The belief that a person can live without consuming food and sometimes also water. Instead, the breatharian body is said to sustain itself on prana, a Hindu life force.

75

In writing this book, this chapter surprised me with its potential for extreme polarization: people with the same environmental end goals take up radically opposed positions, and hold on to them with ideological ferocity. Largely, there seems to be a divide between the traditionalists, and the technologists: the Prophets and the Wizards.

Nowhere is this polarization starker than in our relationship to eating meat. Few argue that we need to eat more meat. Instead the debate tends to move between *less* meat and *no* meat, and the systems implicit behind those positions. Do we want to replace meat with good fakes, or retreat instead from the further industrialization of agriculture? Do we want to tend the wild or grow steaks in labs? Do we need to do it all, and combine the ideal with the pragmatic?

Jonathan Safran Foer attempts to simplify the narrative in his compelling book, *We Are the Weather: Saving the World Begins at Breakfast.* He advocates avoiding factory farming, and eating 'no animal products before dinner', citing research which suggests that an average two-thirds vegan diet is less environmentally impactful than a full-time vegetarian diet.

Safran Foer compares the ease – and significance – of the choice to eat less animal to the war effort in which all Americans partook during the Second World War, to turn off their lights at night. Even in areas where there were no chances of enemy planes, the collective act created connection to the bigger effort.

Researching this chapter, meeting the diverse people working in this space, has offered me great optimism that we are on the brink of a revolution in food. If we can combine these disparate ideas, then we may be able to tackle multiple issues – from planetary to human health – all at the same time.

Perhaps – hopefully – Simon Amstell will emerge a Prophet, and one day the concept of factory farming will be consigned to therapists' chairs! We could reverse the current trend to deforest land for agriculture and instead free up large areas for agro-ecology, or re-wilding. This also offers another possibility: the chance for us to re-engage with the natural world in new, more profound, and wilder ways, and maybe even get our fingernails dirty.

Chapter Three
Resurrection

Plastic, waste and its afterlife

When Maria Carter was asked by her school to write about an environmental solution, she handed in something surprising. Whilst most kids wrote about taking shorter showers or using energy-efficient light bulbs, Maria submitted advice on knife sharpening.

Maria's father, Murray Carter, is a seventeenth-generation Yoshimoto bladesmith. Originally Canadian, he spent eighteen years in Japan learning the ancient craft of smelting and forge-welding steel blades, and carving wooden handles. Eventually he was asked by his teacher Mr Sakemoto-sensei to become the first non-Japanese Yoshimoto bladesmith.

Maria explained that it takes a lot of resources to make a knife – steel from iron ore, coal furnaces, wood or plastic – but when they become dull, they often end up in landfills where those plastic handles will take centuries to biodegrade. Yet as Maria points out, 'Most knives can be sharpened (and reused for decades) until the blade disappears!' Buying one great knife, and learning to sharpen it – maybe even to love it – offers a radically conscious act.

Could you learn to love a knife?

Wasted

What can an individual do to help minimize their impact on the planet? I asked legendary natural scientist David Attenborough. After a pause for thought, his eyes crinkled and lit up: 'Stop waste! Don't waste food. Don't waste power. They are precious, and they are what we require, but not only us. We are taking parts of the Earth on which we all depend.'

In his nineties, David Attenborough is close in age to what my grandparents would be now, were they still alive. He was born during the 1920s dawn of mass-production, lived through the Second World War and its aftermath, when food was rationed and sharing mandated, into an age of globalization and mass consumption, which since the 1960s has seen plastic production multiply twenty times over.

David is now witnessing those changes come full circle: helping to spearhead a global countermovement to reduce plastic waste. Perhaps – hopefully – the roaring twenties of this century will look and sound very different from the last.

No material waste has had as much air time in recent years as plastic, a fair target of environmental outrage. Insanely cheap and prolific, plastics perpetuate our dependence on oil (making up 5 per cent of global demand), contain toxic materials, and take so long to degrade we can't even be sure *how long* – scientists estimate those plastic knife handles will still be around in 400 to 1,000 years.

The annoying thing about plastic (and much other) waste is that there is no *away*. A lot of it ends up being shipped to low-income countries, sacrifice zones where up to a million people a year are understood to die from illnesses caused by waste sites. Of the estimated 50 million tons of electronic waste produced each year, the <u>Environmental Protection Agency</u> (EPA) calculates that only 15 to 20 per cent of it is recycled – the rest is sent to be sorted in US and UK prisons or countries including Ghana, Nigeria and China.

Of the 300 million tons of plastic we produce every year, half is single-use and more than 8 million tons end up in the ocean. When the National Oceanic and Atmospheric Administration (NOAA) first discovered a giant floating island of plastic in the ocean, it remarked, 'It's kind of like a giant toilet that doesn't flush.'

David Attenborough's groundbreaking show *Blue Planet II* brought international attention to the devastating impacts that plastic waste can have on our oceans: polluting, or even killing, the birds and sealife that ingest or get caught up in it: 92 per cent of dead seabirds were found to have ingested plastic; while micro-plastics have been found in 73 per cent of deep-sea fish, and in the faeces of people around the world. Shit, right? If we don't change our consumption patterns, marine scientists estimate that by 2050 there will be more plastic than fish in the ocean.

In the face of growing public unrest within the United States around environmental issues, the EPA was established in 1970 by then President Richard Nixon with the aim to create 'a cleaner, healthier environment for the American people'. Evidence that protest can be effective.

Imperfect steps

We don't need a handful of people doing zero
waste perfectly. We need millions of people
doing it imperfectly.
Anne-Marie Bonneau,
www. zerowastechef.com/

The first and most obvious thing to ask is: can we
waste less?

Some voluntary opportunities to reduce plastic
waste in our lives are simple: stop buying single-use
water bottles; use Tupperware boxes, local farmers'
markets, paper or cotton bags. Some touch points are
less obvious, such as using a Mooncup rather than
sanitary pads.

Yet it's hard to avoid plastic outright, and often it
inserts itself involuntarily into our lives. Food packaging
and kids' products are the areas I regularly battle with.

It is ironic, to say the least, that we often use a
material which will last up to a thousand years to transport the food and drinks we consume in minutes. I
am a diligent recycler, but am regularly disgusted by
the size of my weekly 'recycling' bin.

Meanwhile, the new levels of plastic that enter your
life as a parent are fucked up. An endless deluge of
plastic toys, often given by well-meaning friends or
family; non-biodegradable wet wipes and nappies.
These baby-plastics offer convenience and joy to us
and our children in the short term, at their inevitable
long-term cost.

My daughter oscillates between trying to persuade me to buy her the plastic junk we pass in shops, and helping me pick plastic debris out of the sand in beaches, narrated by her sweet wisdom that it will 'choke' or even 'kill' the turtles (sincere thanks, Attenborough).

Then there is all the invisible plastic: the packaging in global supply chains; the hardware, wires and data centres that allow emails to be sent; the waste in the production process itself. For example, a pair of sunglasses will typically be cut out of a large sheet of acetate, with the offcuts thrown away. In 2017 I helped a friend set up a glasses company (Wires Glasses), which was able to use less than half the plastic that conventional eyewear uses, by 3D-printing the rims instead. It now uses bio-plastics.

Bio-plastics are derived from plants rather than petroleum, meaning they absorb rather than release carbon dioxide. Wires Glasses use a bio-plastic made from the bright pink castor bean plant, which was a favourite of the early Egyptians (for lamp oil) and Thomas Jefferson (believed to deter moles.)

> What most people see in their garbage cans is just the tip of the material iceberg; the product itself contains on average only 5 per cent of the raw materials involved in the process of making and delivering it.
>
> William McDonough and Michael Braungart, *Cradle to Cradle*

We need consumer choices to inspire broader policies and corporate commitments that re-steer and re-shape the landscape. This seems to be under way: in 2018, 'single-use' was the *Oxford English Dictionary*'s word of the year, and there has been a seismic shift in the public consciousness around plastic that is becoming increasingly political.

Many countries (including China, Bangladesh, Kenya, Italy) have banned or taxed single-use plastic bags, water bottles or micro-beads; plastic-free aisles have opened in supermarkets; and companies (including Iceland and Lego) and events (Glastonbury festival and the world's largest concert promoter, Live Nation) have made bold commitments to avoid plastic. The European Parliament, discovering that marine litter (80 per cent of which is plastic) costs them $295 million to $793 million a year, have passed legislation that will mean by 2030 all plastic packaging must be reusable or recyclable.

Meanwhile, reducing energy waste comes up again and again as an environmental policy priority. President Macron's adviser, Aurélien Lechevallier, tells me that improving energy efficiency is 'the most efficient way to reduce our carbon emissions' and 'the most important area in our current situation'.

New laws attempt to push against planned obsolescence. Seeking to 'deliberately reduce the lifespan of a product to increase the rate of replacement' has been a crime in France since 2015. In 2016 France also enacted the world's first law against food waste, making it illegal for supermarkets to destroy or throw away products approaching their sell-by dates, requiring instead that they donate the food, or compost it into methane fuel or animal feed.

Cradle to Cradle is a design philosophy (and book) whereby the components of a product are reused in a continuous loop. Biological material is returned to the biosphere and non-biological materials are repurposed. In theory, this cycle can be repeated infinitely, thus greatly reducing environmental impact. A play on 'cradle to grave', it indicates that materials can reincarnate from one generation to the next.

This law arose after a campaign to celebrate 'inglorious fruits and vegetables', when a French supermarket devoted an aisle to disfigured fruits and vegetables. The initiative was so popular it was expanded and copied around the world, such as the *Fruta Feia* – 'ugly fruit' – cooperative which buys up disfigured fruit to sell.

The rising tide of legislation against different forms of waste reflects a burgeoning counter-culture and demonstrates the delicately webbed interplay between different public actors – politicians, companies and individuals – and how small consumer choices can provoke wider shifts in policy and corporate practice.

Reducing waste offers a win-win solution. Better home insulation means lower energy bills. Meanwhile, in the UK the average household loses £470 a year because of avoidable food waste. Fast fashion may appear cheap, but it's a false economy when you consider how much more you will likely buy. A handmade knife may cost more than the plastic one, but if it lasts a lifetime – even becoming a keepsake to pass on to your child – it offers better value.

Alchemy?

The Stone Age did not end because the world
ran out of stones, and the Oil Age will not end
because we run out of oil.
Don Huberts, *The Economist*,
'The end of the Oil Age'

In spite of all the bans, it seems unlikely that plastic will disappear altogether anytime soon. It is just too useful: it saves lives in medical care, and sometimes you simply want to grab a coffee on the run and the sweet but jobsworthy barista insists on adding a lid for (short-term) health and safety . . . So, can we design better lids and hospital I V bags?

Numerous bio-based alternatives to plastic have already been developed, using algae biomass, corn starch, wood or sugar cane. These developments should be welcomed but bio-based plastics meet criticism because they require land to be grown on, competing once again for space with wildlife. Can we do better still?

I arrived early at a café in San Francisco to meet Josh Hoffman, the CEO of biotechnology company Zymergen. I was killing time, reading an article on queer ecology by Timothy Morton. The logic goes that just as gender is fluid – with no neat separation between male and female – so life is liquid, with no neat separation between us and 'Nature'. If, on a biological level, everything is one continuum of matter transforming into another – there is no such thing as waste, or 'away'.

> Nothing exists independently, and nothing comes from nothing. This is literally about realizing where your waste goes.
> Timothy Morton, *Nothing: Three Inquiries in Buddhism*

Josh arrived and as he described Zymergen's mission to produce new materials and products from biology, I realized he was, in many ways, enacting the thesis of queer ecology, changing one material into another.

According to Josh, the discovery and rise of fossil fuels can be credited for making 1870 to 1970 a 'magical century' that saw enormous advances in human welfare. Our species started to get very good at taking 'basically really old fermented ferns from a million years ago' (aka fossil fuels) and turning them into 'everything': the fertilizer that made the avocados on my plate, the coating on my knife, the power lighting the room, the gout medicine Josh was digesting, the sunscreen on my face, the plastics in my headphones that were sitting on the table.

Ethylene, propylene, butylenes, benzene, toluene, xylenes – these six basic petrochemicals are the Lego bricks that make up such an overwhelming number of man-made products, including actual Lego bricks (although Lego have started making bricks from sugar cane).

'Everything is <u>petrochemical</u>,' Josh insists. 'EVERYTHING!'

Which brings us to a peculiarly twenty-first century dilemma: 'The petrochemical industry has on one hand been this huge contributor to human welfare, and on the other hand, it's killing the planet. You can't really decouple them. Surely, I don't have to pick one of those two?' Josh asked. He is attempting not to.

Nature is full of microbes that magically convert one substance into another: doing what an alchemist could only dream of. How these processes work has long been beyond our comprehension, but Josh (and others) believe that harnessing the powers of biology and machine learning can enable us to crack nature's codes. Josh says with a sense of excited optimism, 'We are biofacturers, we make the building blocks of tomorrow. Fundamentally changing how things are made and what they are made of.' Is this alchemy or Frankenstein?

Josh and his 750-strong team at Zymergen believe they are part of a wave of scientists and entrepreneurs who are heralding in a new age of materials and science, that will allow us to create alternative materials – such as a version of plastic, fuels, fertilizers – without using fossil fuels.

'Biology gives us a far broader, more flexible palette of materials for the world,' he says. 'By unlocking the power of biology, this technology revolution will change everything: what you wear, how you live, what you eat, your health, your well-being, your everything. It's not a promise of a utopian future, it's a glimpse into an astounding present.'

To paint a somewhat clearer picture of this sci-fi reality, Zymergen are working with and improving microbes ('bugs'), under the supreme guidance of their technology platform ('AI plus robots'), to eat carbons ('sugars'), and turn them into new materials ('plastics, adhesives, etc.'). Zymergen are working with a number of large global electronics companies to produce new films for their phones, and other new materials that can replace plastic. They have produced a new form of mosquito repellent, less toxic than Deet, and are working on pesticides. The possibilities, Josh maintains, are 'kind of infinite'.

Apparently, microbes have similar taste buds to children: 'A sugar slurry is delicious for bugs, they want to eat it.' So, right now, Zymergen's are eating sugar. Which poses a similar problem to bio-plastics: in the event of wild success, where to grow lots of sugar cane?

Josh's answer to this is even more radical: 'Now imagine that instead of having to grow sugar for feedstock, I could recycle these plastic headphones and the bugs could just eat them, and convert them into something useful and interesting.' Josh is confident that one day they will be able to close the loop, and make a truly circular supply chain: their microbes will not only produce 'new materials better than plastic', but also consume them.

This is the ultimate vision of recycling, and it isn't wishful thinking: indeed, the science is already on its way. In 2016 Japanese researchers discovered that a species of bacteria had naturally evolved at a waste dump in Japan over the last seventy years, to be able to break down (and eat) the bonds of one of the world's most common plastics, PET/polyester. A process that would normally take hundreds of years took a matter of weeks.

'I think we are seeing how nature can surprise us and in the end the resiliency of nature itself,' said Dr Tracy Mincer, a researcher in ocean plastics.

British scientists then developed this bacterium into a mutant enzyme that begins to break down plastic within days. They believe it will be able to turn plastic back into its original components, enabling far more effective recycling, and reducing the demand for oil and virgin plastic production.

This represents an important shift considering that only 9 per cent of all plastics produced between 1950 and 2015 were actually recycled, and still today many types of plastic (such as cling-film, soft plastics or plastic-lined products like Tetra Pak – depending where you live) are doomed to an afterlife in landfill, incinerators or oceans.

'It's going to happen gradually over time,' says Josh, 'but I believe progress is possible in part through technical innovation. I don't want to expect the oil companies to change strategy, I want to make it easy for them to die.'

Another example of the potential perils of good intentions: conscientious vegans often use much more Tetra Pak for their funky milks. Combining plastic, aluminium and paper in one surface, Tetra Pak claims to be recyclable, but most ends up incinerated. It is, of course, possible to make your own funky milks, and oat apparently has the lightest footprint.

Resurrection

Waste Equals Food.
William McDonough and Michael
Braungart, *Cradle to Cradle*

Is waste in fact an opportunity? What else could we do with it?

When the Johnson Publishing Company (JPC) – once the largest African-American-owned publishing company in the United States – went bankrupt in 2019, artist Theaster Gates wondered if he could recycle it. 'I started to ask myself, could an artist do something with the corporation that a business person wouldn't be willing to? Could I get people excited about this seemingly bankrupt thing by changing the form?'

Theaster licensed the JPC images for exhibition, shedding light on the beautiful black women of the 1950–60s that JPC's magazines, like *JET* and *EBONY*, had originally celebrated. He also took over the building: ripping out the carpets and furniture to re-upholster and re-deploy them in new spaces; he took JPC's collection of 26,000 books and created a library inside the Stony Island Arts Bank (a building he had already converted from a black-owned Chicago financial institution).

'We imagine that all of these things have no value, simply because people haven't taken the time to dig deep and understand the possibilities,' says Theaster. 'I definitely believe in the power of resurrection: that is, that a thing could be seemingly dead and that through belief and hard work, one could find oneself interrupting the death cycle to bring life again.'

Theaster Gates, 'The Black Image Corporation', Fondazione Prada Osservatorio, Milan, 2018.

All around the world, 'waste' is being resurrected in new and creative ways. The millions of tons of pineapple leaves, orange and apple peel, fish skins and grape skins produced by the food and drinks industry, and even cow faeces, are being used to create fabrics.

In 2017, I wore a silver dress by punk queen Vivienne Westwood, made of recycled plastic bottles, to the Oscars.

Thread International have paid local people in Haiti and Honduras to collect millions of plastic bottles from the streets, canals and landfills, which are then recycled, spun and woven into fabrics. 'If business can figure out how to use the $500 billion worth of waste that the fashion industry leaves behind every year, then not only are they saving the planet, they also have a very healthy business model,' Stella McCartney told me.

Electronic waste offers a unique opportunity for recycling, as valuable raw materials, finite minerals and metals can be reused and re-purposed. As we saw in the last chapter, food waste is being used for regenerative agriculture, increasing the soil's capacity to lock in carbon. <u>Human faeces</u> is being used to generate electricity, and household waste to fly planes!

UN researchers say that bricks of human poo, made in kilns, would have the same energy content as coal.

Looked at with these eyes, waste can be re-appraised as an opportunity for creating something new. Remember, nature doesn't do waste: energy can only be transformed, not destroyed. The question is what we want our energy – our waste – to be transformed into.

Junking the waste mindset

waste **v. 1** use carelessly, extravagantly, or to
no purpose.
Concise Oxford English Dictionary

The decision to waste, or the effort to resurrect, comes
from a cultural mindset. For thousands of years,
human communities actively discouraged waste. It
was always uneconomic, impractical, and even sometimes viewed as spiritually negligent. For nomadic
communities, waste was a burden to be carried, and
should be avoided at all costs. If a San hunter, in Botswana, killed an animal too large for the community
to eat, he would be heavily criticized.

Where my mother, Patience, grew up, on a very remote farm at the top of a mountain in South Wales, as
one of three children, there wasn't a rubbish collection.
This was partially because there was no road to the house
– it required a three-mile walk through grassland – but
also because there was no real need for it: self-sufficient
in terms of most of their food needs, and without much
money to buy extras, they had very little rubbish.

'Everything possible was recycled!' my mother says.
Her parents, Sylvia and Wyndham, used sheets of cor-
rugated metal to build a new kitchen, and, 'When we
came home from school we had to keep our chewing
gum so my mother could push it up into the nail holes,
to stop it from dripping.'

Almost nothing at that time and place could be
thrown away. My father's brother fished out and kept

a shoe he found in the local river. When asked what use was one shoe, he replied, 'It's a bloody good shoe!'

Sylvia handmade most of the girls' clothes. My mother was given one blouse or dress to wear all week, and was mortified to be sent to school in her grandmother's old shoes. Still today, I rely on my mum to help me darn and fix tears in my clothes; and watch her endlessly reuse her teabags, bottles and carrier-bags.

I would not swap my life to be amongst the San a hundred years ago, or have my mum's childhood in poor rural Wales. There is a lot about our culture that I enjoy: hot baths and crisps. Yet I do think we ought to look to these communities of the past, and the realities of the present, to see what we might learn, so that we might re-route to find more sustainable ways of being and thinking.

Our disposal culture represents a very new form of decadence, when products appear cheaper than they really are, and the pursuit of convenience reigns supreme. It also reflects an unbalanced world: half of its population live on less than $5.50 a day; and the average person in the US or Australia has a carbon footprint one hundred and fifty times larger than someone in Ethiopia. (Within affluent countries the disparities widen again: such that someone from Wyoming has on average a thousand times the footprint of the average Ethiopian.)

It is a painful paradox of our age that we live in a time of great abundance, yet because of inequality and waste, many still go without. In affluent societies, our waste offerings are the dead body, the carcass, the evidence, of our consumption.

The focus in this chapter has been on plastics, as that is where we are most readily seeing innovation in the move away from single-use petroleum-based materials, but plastic is just a metaphor for everything else: food wasted because it is ugly or close to its sell-by date; energy wasted because of bad insulation; all the little things that fill the space: paperclips and mobile phones; diaries and the ink that writes in them; bank statements and legal paperwork; old hard drives, USBs, CDs and outdated charging cables; printed photographs and broken plant pots.

David Attenborough's answer to that big question lingered with me: philosophical and all-encompassing. More than innovation or regulation, what he seemed to be proposing was a fundamentally cultural shift. Can we move away from overconsumption and disposability? Can we work against capitalism's insistence that new is better? Can we take personal responsibility for our own small choices, and redefine our personal and collective relationship to the material world? Can we know the value in a special dress, ugly fruit, or beautiful knife asking to be fixed, repaired, shared, sharpened, loved?

PART TWO

Magic Tech Bullets

Will technology save the day?

Any sufficiently advanced technology is indistinguishable from magic.

Arthur C. Clarke

I once spent several months on the abandoned rooftops of crumbling factories, filming a dystopian
art-house movie. Civilization had collapsed and
humans had returned to a more primitive survivalist
state: cannibalism and violence were interwoven with
mysticism and early human biblical mythology. The
director had originally planned to film it in a desert,
but then he stumbled on Detroit.

Contemporary Detroit offers up a surreal post-
technological, post-apocalyptic landscape. A city that
rose around the success of the automotive industry
has gone through a consequent cycle of economic
decline, and miles of giant factories and monolithic
buildings have been left to ruin. Concrete beams lie
like the fallen pillars of another era, and trees have
reclaimed the spaces in between.

A resurgent artistic counter culture has emerged
in Detroit: when I was there houses were priced at a
few hundred dollars and artists had bought up whole
streets to play with. They lived communally and ran-
sacked supermarket waste bins for dinner.

In the centre of the wasteland, large murals painted
by Diego Rivera in 1932–33 offer a preserved image of

the industrialism that was – at one moment in time – the heart of Detroit. Rivera spent eight months there, studying the city's web of productivity. In the twenty-seven panels he painted, mechanization and technology take on spiritual and allegorical dimensions: a large anthropomorphic stamping machine is a reference to the Aztec goddess Coatlicue; a Christ-like child is being given a vaccine by a doctor who appears to be Joseph and a nurse Mary; in the background three scientists – the biblical Magi – are conducting research. Put simply, technology is the new religion.

In the 1920s and 1930s, Detroit was at the centre of a technological revolution. The Ford Motor Company had pioneered the automated assembly line, leading to the mass-production of cars that made them affordable to millions of people, and the city had diverse factories producing all the composite parts needed also for ships and aeroplanes.

Rivera's murals also captured the advances in science and medicine that were shaking Detroit and the wider world. Although some have interpreted his work as more parody than allegory, especially given his own Marxist political leanings, the museum maintained that Rivera 'thought our mass-production industries and our technology wonderful and very exciting, [and] painted them as one of the great achievements of the twentieth century'.

There seem to be two versions of dystopian futures painted and imagined by sci-fi film writers and directors. One offers a primitive return to nature, after the collapse of technology – think *Mad Max* – and the other imagines technology taking over – think *Blade Runner 2049* or *Black Mirror*. When we consider the percolations of where technology goes next it is easy to be dystopian: augmented-reality advertising targeted at individuals based on their biases; mass surveillance of every call or text we make; cyborgs; Artificial Intelligence a billion times smarter than us driving robots in war.

Yet, perhaps ironically, technology is also, and simultaneously, still our religious Coatlicue: looked to by many hopeful people as the potential saviour of our species, and panacea to our lifestyles. Maybe, as cars get cleaner, renewables get more funding, electric planes take flight, labs cook up 'clean meat' and scientists invent carbon-capture devices, we can afford to relax, and not question our behaviour too rigorously?

Maybe.

Yet that is arguably a high-risk strategy to bank one's survival on, breeding complacency. If you are buying the tech-will-save-us narrative, first you must pause to consider the plumbing behind technology's 'magic'. What impact is digital really having?

Chapter Four
Rainbows and Unicorns

Smartphones, digital and the cost of free

I was standing in the 'Valley of Heart's Delight' – or Silicon Valley as it is better known – by an artificial pond, watching a lone duck drift across lines of colour, drawn in the water's surface. It was late, and dark, but the sky around us was alive with colour. Icelandic artists Olafur Eliasson and Jónsi (from avant-rock band Sigur Rós) had brought the aurora borealis and trance-like waves of sound to California for one night. I wondered what the duck thought.

The pond, duck, Al Gore and I were amongst oaks, apricot and apple orchards, at the centre of the Ring: a circular building a mile wide, which hosts one of the largest roof installations of solar panels in the world. A gigantic rainbow sculpture rose up nearby.

By my side was one of the minds behind this feat – Stefan Behling, studio head from Foster & Partners, and expert in sustainable architecture. He was buzzing excitedly, watching the opening of this ambitious new building, the culmination of ten years' work. No less than a park. Nicknamed the 'spaceship' and the 'mother-ship', the circle a hallmark of utopian architecture, it certainly felt like a <u>reality distortion field</u>.

Building the future is expensive. Who else could be behind this moment, but the world's largest

Used to describe the phenomenon whereby an individual is able to make others believe in the possibility of achieving very difficult – or near impossible – things. The term is borrowed from an episode of *Star Trek* and was commonly used to describe the motivational talents of late Apple CEO Steve Jobs.

technology company? Apple Park is a 175-acre site that is 80 per cent green, powered by 100 per cent renewable energy and contains the largest naturally ventilated building in the world. The site was previously 80 per cent covered in buildings and car parks: 'You might as well have put a plastic bag over the earth,' Stefan told me. 'We had to rip it all up, and let the earth come out and breathe.' They recycled nearly all of the existing construction materials.

Inside the Ring, large sheets of curved glass create windows from which to see 9,000 indigenous and drought-resistant trees. There are no cars – people walk or cycle to navigate the undulating hills. 'The magic of Apple Park is the blending of the inside space and the outside landscape. It is deeply connected with the ideas of biophilia and human beings wanting to be connected to nature.'

The flora and fauna will capture approximately 200 tons of carbon dioxide annually: seeking to make the project 'climate restorative', a new architectural buzzword for developments that attempt to give back more than they take. Although Stefan points out that the next generation of buildings will need to go even further, and also need to consider the total embodied energy used in construction.

Close by, across San Francisco giant billboards bring images of giraffes and sharks into the urban landscape: advertising Apple's phone cameras, but also subtly instating their shift towards sustainability. Is it all green-rainbow wash? Or is one of the largest companies in the world changing the rules and expectations for technology's footprint? Should we be hopeful that this really is the future?

The hard cloud

Smartphones have out-smarted us. These magical objects enable us to navigate ourselves around the world, take our office with us wherever we go, stay in touch with loved ones, and access libraries of information. They are even starting to impact our health and fitness through wearable devices. They help farmers track weather conditions, and empower entrepreneurs, and people without bank accounts, through mobile banking. They enable us to be politically engaged and informed, and to organize grass-roots campaigns, whilst a proliferation of applications offers everything from entertainment to fertility calendars. Phones have become so intelligent, even addictive, that sometimes it feels like we can't live without them. Can you?

People often think of the Internet as effervescent – we send a message and it instantaneously arrives on the other side of the world, as if by telepathy. The 'cloud' denotes images of airy constellations in the sky. We forget, or do not know, that the cloud and its digital landscape depend on many metals, minerals and plastics, all of which come from somewhere, and require eye-watering amounts of energy – usually in the form of server farms in the Arctic – to keep running.

For every phone in someone's hands, laptop in a bag, TV at home or electric car in your dreams, there is a little hole in the ground somewhere: cobalt mines in the Congo, gold dug in Australia, tungsten in China, nickel from the Philippines. If you go to the Democratic Republic of Congo and spend time in the Great Lakes Region, where most of the world's cobalt is mined, you'll find both large and artisanal small-scale mining perpetuating a decades-long armed conflict that turns children into soldiers.

> I want people who don't get to see uranium
> and cobalt to realize that it's here. Because
> I see it every day. The Congo is here. We just
> have to wake up.
>
> Freddy Tsimba, *Mavambu* (film)

Electronics is perhaps the final frontier in 'conscious consumerism'. Leading the way is a small Dutch company called Fairphone, which began in 2010 as a campaign centred on creating a conflict-free phone. Their CEO, Bas van Abel, wanted to disrupt the electronics sector and likes to tell the story of how it could all have started as a (fair) toaster, but instead they decided to tackle the mobile market's poor social and ecological record.

By 2012, the campaign had developed a community of 4,000 potential customers, and had built a network of relationships with NGOs, operators, phone producers and the Dutch government envoys on conflict minerals who shared this vision. They launched a very successful crowd-funding campaign to build their first phone, and managed to go into production of the first Fairphone with 35,000 pre-orders.

Fairphone sought to have impact in the electronic supply chain in four main areas: sourcing conflict-free minerals; looking at workers' welfare in the manufacturing process; offering a modular design which encourages users to fix their phones rather than replacing them annually; and considering the end of life of the phone to try and make it more circular.

'The phone is an amazing symbol of how our economic systems work, because it's not like a banana or coffee where there is just one supply chain – there are thousands of supply chains and no one really knows how it comes together. How can we source fairer minerals? How can we disassemble a phone and make the supply chains around it more transparent?' said Tessa Wernick, founding member of Fairphone.

Fairphone wanted to support African miners by encouraging a conflict-free industry in the Congo, so they joined existing initiatives on the ground, and set up their own in Rwanda. Consequently, they managed to trace the four minerals most at risk of coming from a source of conflict, from conflict-free sources.

'Phones are really closed systems – people have no idea where they come from,' says Tessa. 'Some partners, including other suppliers in the supply chains, never knew about conflict minerals. We would go to them with information so that they could make those decisions with more transparency.'

Fairphone provoked a conversation which has been felt across the industry, but Tessa is honest about their struggles and limitations: as a small company they don't have the same bargaining power with suppliers that a large one might. If Fairphone's mission was to be a thorn in the foot of the technology industry, they have succeeded. Now other, larger players are investing billions of dollars in cleaning up their supply chains, addressing both social and environmental concerns.

The big Apple

Apple's business model is fairly simple. They sell products. They are like a big, fancy version of a shoeshop (albeit one that makes shoes from underground minerals, and a hell of a lot of them). However, I always struggled to make peace with the environmental, and social, cost of their insistence that we need to buy new products every year (and new chargers, adapters, cases and headphones to fit them).

To their credit, Apple have – at this critical moment in time – hired one of the best people in the world to try to square this circle. Coming from the EPA, where she was administrator from 2009 to 2013 for President Obama, Lisa Jackson is experienced and nothing short of ambitious: 'I like big stretch goals which really require you to dream,' she tells me (on an iPhone call) on her way to work.

Apple have been recognized as a leader in sourcing conflict-free minerals, and now are focused on addressing their environmental footprint. In a letter to the EPA, they stated that the fight against climate change is a 'moral and environmental imperative that also makes good business sense'.

'I think our customers really expect us to tackle the hard problems,' says Lisa of their new initiatives. She set the company a bold, essential challenge by asking, 'Can we one day stop mining the Earth altogether?'

Though Apple haven't set a date for this goal, they are on the path towards it: they now offer consumer buy-backs of products; have created recycling robots 'Liam' and 'Daisy'; and have increased the amount of recycled plastic, tin and aluminium in their products.

Lisa has also overseen the company's move to using 100 per cent renewable energy in all their global facilities (retail stores, offices, data centres), with the goal of getting all their manufacturing suppliers to move to the same in the near future. This announce-ment came in 2017, just a week after Google had announced it was using 100 per cent renewable energy for its operations.

After spending twenty years at the EPA, Lisa believes that the private sector will be at the helm of change: 'Inside the EPA, you're never making a regulation that's so cutting edge and out there that it changes everything. What we tend to do is make regulations that shore up the bottom. In terms of leading from the front of the pack – the people who can say, OK, here's a whole new way to think about this problem – that's going to come from the private sector, and also from research.'

Apple has not been without its share of criticism over the years: in fact, there is a long Wikipedia entry devoted to 'Criticism of Apple Inc.'

From an environmental perspective, their business model presents fundamental contradictions: encouraging people to regularly buy more technological products inevitably contributes to the 50 million tons of electronic waste produced every year, much of which ends up polluting the lower-income countries it is shipped to.

Apple are also being investigated by a French consumer protection agency, after the slowing down of older phone models prompted accusations of planned obsolescence. Apple argue that this slower software performance is a result of the batteries degrading, and in response have reduced the fees for their battery repair services.

The desire to repair

Planned obsolescence has been a central component of techno-capitalism ever since the phrase was first coined in the 1930s, in response to a strategy announced by the head of General Motors: to make annual design changes to the cars, to convince owners that they needed to buy new models. The strategy eventually went beyond design changes, with companies engineering artificial lifespans, so that customers would be required to buy the products again.

My mum likes to tell the story of her friend Penny who worked in a sewing-machine shop in England in the 1970s. The shop offered people a discount on new machines if they brought in their old ones: shelves at the back of the shop were apparently filled with beautiful original machines, ready to be sent off to be destroyed, because 'they were made to last, preventing new sales'.

Planned obsolescence arguably exists in all industries, but it has managed to take root in technology in a deeper way because the complexity of the products, and the quick pace of innovation, often disempowers the consumer from understanding how long things should be expected to last, or knowing how to fix or update them themselves. Light bulbs and printer ink cartridges are infamous prey to this cynical twentieth century strategy.

The 'Right to Repair' movement in the US began with farmers who wanted to be able to fix their high-tech machinery, but couldn't find service information or spare parts. It was successful in the automobile industry, when in 2012 Massachusetts passed a bill requiring automobile companies to provide the necessary documents to allow anyone to repair their vehicle. There are currently numerous similar US bills aimed at electronics manufacturers, asking for repair information and spare parts to be made available.

In 2019, the European Union announced new laws that mandate that large electronic items (such as washing machines, TVs and fridges) are designed so that spare parts can be available to licensed third-party professionals to repair products, and extend their lifetime. These laws indicate a trend that is likely to accelerate.

Apple have created enemies by lobbying against Right to Repair bills – although they claim this is because of security risks. Nathan Proctor, involved in the Right to Repair effort, says, 'The debate isn't really about how green [Apple's] new products are. It's about how they treat the older devices. They should give people better tools to keep older devices going.'

Planned obsolescence is a more obvious and outrageous manifestation of a deeper, ongoing narrative of our prevailing systems: the obsession with economic growth. How do you keep financially growing in a consumer society? Is selling more products ever going to be materially sustainable? Perhaps if you can design completely recycled, closed-loop products, made with 100 per cent clean energy, that last a long time. Which is what – you might recall – Lisa Jackson is helping to get Apple doing.

Lisa has a hard, if enviable, job. Enviable because she has the opportunity to have a huge environmental impact by guiding one of the largest companies in the world towards better working practices, and demonstrating what is possible to the global community. A hard job because of the inevitable tensions that she is being asked to reconcile: how do you make a company that is legally answerable to thousands of anonymous shareholders' bottom lines, and whose business model is fundamentally based on *people buying new stuff*, environmentally sustainable?

Others are taking a different approach, and along-side the rise of people trying to fix their own phones, or buying into small players like Fairphone, there has been a resurgence of those stepping away from smart-phones altogether by buying 'dumb-phones' that only allow texting and calling: in 2017, the sales of 'dumb-phones' grew 5 per cent, compared to that of 2 per cent for smartphones. Some are doing away with hav-ing a mobile phone altogether.

After I lost my phone in 2018, I decided not to replace it. I managed without it for a few months be-fore I borrowed one for a work trip and was quickly sucked back in. As much as I appreciate the benefits of having a smartphone, I actually really loved doing without. My mind was invited to wander whenever I had a spare second rather than reach into my bag. I did get lost a few times, took fewer photos and once struggled to find a friend I was meeting – but it really didn't shake up my life all that much.

When asked about the possibility of sustainable growth, and the paradox of encouraging people to buy more alongside being environmental, Lisa talks about reality and desire: 'I think I should be real-istic – too often environmentalists or sustainability advocates are asking people to work against human instincts, like the human desire for pretty things, the human desire for something nice, the human desire to have a home of your own, the human desire to maybe have a car.'

'We can't always be the ones saying you shouldn't have. What we have to do instead is insist on making

In 1927, leading Wall Street banker Paul Mazur argued: 'We must shift America from a needs to a desires culture. People must be trained to desire, to want new things, even before the old have been entirely consumed.'

people's *normal* work in a way that leads to sustainability.' Lisa compares this logic to the wider geopolitical debate on slowing development: 'A lot of developing countries say hang on a second, you had your moment and now you want me to give it up? Which is just not going to sell!' And hey, Apple know better than most what sells.

What if we start to *desire* more durable, and circular, products?

Chasing unicorns

I had just (unknowingly) celebrated my first birthday when Tim Berners-Lee filed his proposal for the information management system that was to become the Internet, to which his boss at CERN responded, 'Vague but exciting.' Today, half of the world is online.

The invention of the Internet has metamorphosed our world. It has enabled access to an extraordinary and unprecedented amount of information, connected us to diverse opinions and people, and continually transforms different institutions. Digital gives more reach for us to buy better products, build social movements, communicate our stories and messages, and understand diverse perspectives.

In my last year of university, I had an idea to use technology to connect people to trade in new ways (called Impossible – more on that later), and I enthusiastically dived head first into that world. My rebellious history teacher visited when I was working

Rainbows and Unicorns

on Impossible and his mind was blown by the idea that technology could allow us to try ideas that activists had only been dreaming of for thousands of years: to collectively organize and create new currencies. I was a tech-utopian, and the possibilities seemed limitless. I was not alone in my optimism. There was a moment when many people believed the Web was going to make democracy more meaningful, bring transparency to institutions, obliterate hierarchical power structures, create new open-access media and generally revolutionize our world. Yet, in Silicon Valley, I discovered a morally complicated picture of ideals, chasing 'unicorns'.

In the world of myth and legend, a majestic horse with a single horn on its forehead. In the fantastical world of venture capital, a privately held start-up valued at $1 billion or more.

> Some days, lying on the floor next to my one-
> year-old son as he plays with his dinosaurs,
> I catch myself scrolling through Instagram,
> waiting to see if the next image will be more
> beautiful than the last. What am I doing?
> Facebook co-founder Chris Hughes,
> *New York Times*, 'It's time to break up
> Facebook'

I was an 'early adopter' of one of tech's first high-profile unicorns. I had always liked taking photographs, but the combination of having a camera in my pocket (i.e. phone), and a community to share photos with, fuelled my enthusiasm. I found myself seeing the world differently: constantly training my eye on things around me, looking for images to capture and share. I really loved Instagram.

Many years later, I noticed a neurotic channel in my brain deliberating about which photo to post, how previous posts had performed, or what I had recently seen on the platform. There was a popularity contest going on in my mind, with hard numbers attached. My relationship to Instagram – and my phone itself – had morphed into an addiction, and like most addictions, it wasn't an entirely rosy picture.

I wasn't alone. Some 40 per cent of the world's population use social media, averaging two hours a day on social and messaging. Whilst it no doubt has positive impacts in our lives, numerous studies also indicate a prevalence of negative psychological effects such as anxiety, loneliness, <u>FOMO</u>, cyber-bullying, polarization, and depression, especially high amongst girls.

That uncanny and unshakeable feeling that something exciting and infinitely rewarding is happening, precisely where you aren't. All your friends are probably going to be there too and they won't stop going on about it when you next see them.

Digital's addictive qualities are perhaps unsurprising. Technology giants employ some of the smartest minds in the world to constantly design and redesign the platforms to literally addict us. They track user numbers and engagement metrics – always striving to drive them up. It's a simple equation: our time is their money.

Big data has become big business. Most of the digital landscape is propped up by sophisticated data analysis machines that can better advertise according to our biases: making us buy things we didn't know we needed; or vote for ideas or parties we weren't aware we believed in. A test on Google's advertising data found that their algorithms identified when women were pregnant even before they knew themselves, from subtle changes in their search behaviour.

Meanwhile technology companies including Facebook have started investing in mind-reading technology.

Whilst most data intelligence has been applied agnostically, as an atomic force for consumption, to nudge us to buy a sofa, or the trip we emailed a friend about earlier – could this intelligence machine be used to further better ends? Some Artificial Intelligence researchers believe so. In the paper 'Tackling Climate Change with Machine Learning' the authors posit that big data can be used to help solve our climate crisis through individual awareness and action: 'Techniques that advertisers have successfully used to target consumers can be used to help us behave in more environmentally aware ways,' they say, citing research showing that these techniques have improved the enrolment into an energy savings programme by two to three times.

Is 'good' manipulation, *good*? In an ideal world, we would *own* our own data, and be able to choose how it is intelligently used to guide us. I would love to know how I could lower my carbon footprint further: but I would love, even more, to *choose* to know.

How susceptible are we? I used to maintain that I could not be affected by advertising: I would never buy a sofa or a holiday simply because a pop-up asks me to. I would simply ignore it. Yet someone once asked me, what if the data knew you cared about the environment, and so you started seeing the green positions of political candidates? What if some of the information <u>wasn't even true</u>?

Fuck. They would get me.

Facebook, at the time of writing, explicitly exempt political adverts from third-party fact checking (but heaven help you if you want to get a hint of a nipple past their censors (#FreeTheNipple).

Is free will an illusion?

I was first introduced to Cambridge Analytica when I was building Impossible, and someone on our team recommended them as a way to promote our product. A Cambridge psychologist had applied psychological analysis to people's patterns of 'likes' and other Facebook data (gender, age, ethnicity, location, friends) to make assumptions about what kind of messaging might successfully influence them. My mind was blown, but it didn't sit comfortably with me and we decided not to use it.

The real dangers of social media's impact came into play with the 2016 US presidential election, and the UK Brexit vote. Studies have found that in 2016, a third of Twitter traffic in the run-up to the EU referendum in the UK came from bots, whilst in the run-up to the US presidential election, almost half of the population were likely to have seen a Russian-linked post.

Cambridge Analytica was exposed, by journalist Carole Cadwalladr and whistle-blower Christopher Wylie, as playing a role in spreading misinformation to influence voters, using the data from 87 million Facebook users which had been accessed without their consent (breaching Facebook terms). Cambridge Analytica claimed to have worked on over 200 elections around the world, although some political scientists remain sceptical of their actual impact.

The data scandal prompted the investigation of Facebook by governments around the world; the company has been fined $5 billion for mishandling user data in the US, and been the subject of a lawsuit from Canadian regulators.

In 2019 former President Obama conceded that the ability of social media to 'scramble facts' and create polarizing information bubbles is the thing that makes him 'worry most right now'. He argued that 'we will have to find ways to build regulatory structures, parameters and mechanisms, whereby these incredible platforms at least are compatible with the core values of our society'.

Obama joins a growing list of individuals and politicians calling for better regulation of tech giants. US Senator Elizabeth Warren is calling for a 'grass-roots movement' to break up big tech. Facebook co-founder Chris Hughes has also called on US regulators to unwind the company's purchase of WhatsApp and Instagram – and thereby enable more consumer choice. It is an idea with precedent: anti-trust lawsuits were brought against AT&T and IBM in the 1980s, and Microsoft in the 1990s.

Active participants in the tech world, from employees to activist investors, have also exercised their power from the inside, to rally against corporate policies they do not agree with. For example, in 2018, after 4,000 employees protested and some even resigned, Google chose not to renew an AI contract with the Pentagon. In 2019, the day before several thousand Amazon employees walked out in a protest against their company's environmental position, their CEO Jeff Bezos announced that the company would use 100 per cent renewable energy by 2030, and have net zero emissions by 2040.

Sometimes voices from the inside are heard.

The cost of 'free'

It is the consumer who is consumed. You are
delivered to the advertiser who is the customer.
He consumes you. You are the end product.
Richard Serra and Carlota Fay Schoolman,
Television Delivers People (film)

In 1973, in a seven-minute video, with yellow text scrolling down a blue screen, and chirpy upbeat canned music playing in the background, artists Richard Serra and Carlota Fay Schoolman delivered a poignant lecture on the economic dynamics of 'free' mass-media television and news: arguing that the audience is the product, to be delivered to the advertiser.

In the decades since that video was made, we have witnessed a proliferation of 'free' services, which deeply complicate the narrative of conscious consumerism. Here, we cannot utilize the political power of our wallets. Yet what is the currency we pay with? Time, attention, information and mental space.

Chris Hughes has singled out the fact that Facebook is 'free' as a central reason why it has not been better regulated to protect consumer interests. This, he argues, has enabled the tech giant to become a monopoly, crushing or buying out all competition. When the company receives 80 per cent of social media traffic (via Facebook, WhatsApp and Instagram) – and their only real competition (Twitter) employs a similar business model – where, except offline, do we turn? Likewise, when Amazon receive nearly half of all e commerce, and 70 per cent of all Internet traffic goes via Facebook and Google, where is the consumer choice that capitalism is usually celebrated for?

To their credit, in 2019, Twitter and TikTok banned political advertising on their platforms. An imperfect solution, but definitely an important step in the right direction.

'The biggest winners' of breaking up Facebook, Hughes argues, 'would be the people. Imagine a competitive market in which they could choose among one network that offered higher privacy standards, another that cost a fee to join but had little advertising and another that would allow users to customize and tweak their feeds as they saw fit. No one knows exactly what Facebook's competitors would offer to differentiate themselves. That's exactly the point.'

Meanwhile the data mined by companies has enabled governments to get unprecedented access to their citizens' digital communications. At enormous personal cost and risk, whistle-blowers such as Edward Snowden have exposed the mass ow—bus

ingov*Magic Tech Bullets*

surveillance programmes adopted by many Western governments.

After working with the CIA, Snowden revealed that the National Security Agency in the US – and their affiliated organizations such as GCHQ in the UK – paid billions of dollars to telecommunications and private tech companies, and tapped underwater cables, and Yahoo and Google data centres, to access the digital communications, telephone calls and texts of millions of ordinary citizens and people of interest, in the US and internationally. German Chancellor Angela Merkel compared the NSA with the Stasi, and another NSA whistle-blower, William Binney, said the expansion of state surveillance is 'better than anything that the KGB, the Stasi, or the Gestapo and SS ever had'.

Adapting the arguments of conscientious objectors who refused to take part in warfare on moral grounds throughout the twentieth century, conscientious protection is the act of refusing to take part in the harming of the natural world and the moral compulsion to take direct action against it. Used increasingly as a legal argument to defend environmental civil disobedience.

This threatens our democracy in a different way by undermining checks and balances on the state's power. If you think, *I have nothing to hide*, you miss the point: that our own laws are liable to change, and they are not always just. Remember, there are still over seventy countries around the world where it is illegal to be gay, sometimes meeting the death penalty. Or if you look back in history, mass citizen movements have often had to resort to civil disobedience to seek positive changes: the campaign for female suffrage; the push to abolish slavery; and the civil rights movement. Today, as we will discuss in Chapter Eight, the environmental movement is resorting to civil disobedience and conscientious protection, to try and enact change.

Following on from Snowden's revelations, a federal court in the US ruled that the NSA programme was unconstitutional, violating the fourth amendment right of 'no unreasonable search and seizure' and in 2013, the United Nations General Assembly unanimously adopted a non-binding declaration of the right of all individuals to online privacy.

'The good news is that there are solutions,' Snowden told the European Parliament. 'The weakness of mass surveillance is that it can very easily be made much more expensive through changes in technical standards.' Snowden recommends moving away from the illusion of *free*: 'Tell customers they can pay $5 a month for <u>encryption</u> – many consumers will want to pay for that.'

Platforms like Spotify offer successful hybrid models (with a free ad-based or paid-for option); Netflix is subscription-based; Wikipedia operates on donations; and the *Guardian* introduced optional membership, whilst other online news platforms have paywalls.

Personally, as I became mindful of the addictive nature of social media platforms, and the wider political ramifications of their misuse of data, I realized that thoughtlessly using these platforms – buying into them with my time – was like eating carcinogenic junk food or factory-farmed meat. Certainly not nutritious, and potentially harmful to society at large. I started to diet.

Of course, there are advantages to digital content being free: it is easier to find material and it makes

Ancient Egyptians, Spartans, Israelites, Romans, Silicon Valley tech wizards; all have understood the importance of transforming sensitive information into incomprehensible gobbledygook, and back again, with the help of a special key, cipher or algorithm.

platforms accessible to poorer communities. Plus, who doesn't enjoy getting something for free? Yet as a society, we need to think deeply about how we are actually paying: what we are giving away as part of that Faustian pact.

The Web We Want

I am aware that, for a book on optimism, this last section has been very negative. Yet our increasing awareness of the challenges – the bravery of whistle-blowers, and the determination of lawyers, journalists and judges to bring truth to the light of day – gives me hope that we can harness and protect a vision of the Internet that is open, democratic, free and respectful of essential civil rights like privacy.

Imagine giving a computer to a cavewoman. The digital world offers mind-bending possibilities. In so many ways it has reshaped our lives for the better. Digital is also helping us, in many ways, <u>dematerialize</u> our lives: streaming digital films and music negates the need for boxes of VHS and CDs. The fact that a smartphone functions as a camera, scanner, map and speaker means we can use one piece of hardware instead of four or five or more.

I am no longer a tech-utopian, but I do still believe in technology's potential. It is worth fighting for.

On the thirtieth anniversary of the Web, Tim Berners-Lee issued a letter galvanizing a global

Ephemeralization was the vision espoused by Buckminster Fuller that technology would enable you to do 'more and more with less and less until eventually you can do everything with nothing'.

citizen-led movement to push companies and governments to create a 'new contract' for the Web We Want.

'Against the backdrop of news stories about how the Web is misused, it's understandable that many people feel afraid and unsure if the Web is really a force for good,' he said. 'But given how much the Web has changed in the past thirty years, it would be defeatist and unimaginative to assume that the Web as we know it can't be changed for the better in the next thirty. If we give up on building a better Web now, then the Web will not have failed us. We will have failed the Web.'

The 'Contract for the Web' seeks to establish a set of norms to guide digital policy and company decisions. It has been backed by over 200 companies including tech giants Google, Facebook, Twitter and Microsoft, 100 civil society organizations, three national governments and thousands of individuals from over 170 countries. It is a document that can be used by citizens and activists to hold large players to account.

The Web Foundation's CEO Adrian Lovett tells me, 'We aren't powerless bystanders and we shouldn't sit on our hands and leave companies and governments to decide what our digital future looks like for us. What we really need is a movement of people who care about these digital issues and who are willing to take action. And that's starting to happen.'

Damian Bradfield – author of *The Trust Manifesto* – offers this advice: 'You and I need to be much more aware of what is happening, stop acting like ventriloquists and actively choose alternatives. Choose DuckDuckGo, for example, for searches. Choose not to use Facebook. Or Instagram. Or if you do use them – fuck with the data, turn sharing off, amend your location to pretend you're in Egypt. As businesses, 87 cents of $1 in marketing goes to Google and Facebook, with Amazon gaining ground. Support other publishers. Lastly there does need to be regulation: Amazon controls 60 per cent of the servers of the Internet! Government needs to break up big tech. I see this as all being very much within reach – like the organic food movement, it's about education and us demanding change.'

Tim Berners-Lee is calling on citizens to lead the way in demanding improvements: 'The Web is for everyone and collectively we hold the power to change it. It won't be easy. But if we dream a little and work a lot, we can get the Web we want.'

Also, worth checking out Ecosia.org who plant trees when you use them for online searches! At the time of writing they had planted over 71 million trees.

Wielding the sword

Staring at the screen so we don't have to see the planet die.
Kate Tempest, 'Tunnel Vision' (song)

If the Internet were a gigantic iceberg, the tip, just visible above

Behind the vision of technology as a world of rainbows and unicorns, there exist contradictions, challenges and shadows. (And I haven't even got into the 'Dark Web.')

Rainbows and Unicorns

We must remember our digital landscape is insanely new: younger than me, it offers a blip in human evolutionary time. Perhaps it just needs to grow up.

My optimism for technology is rooted in the fact that electronics giants are starting to clean up their supply chains, boost recycling initiatives, and build more circular products. Hopefully, in time, this shift will start to impact other electronics areas: what about a fair washing machine? A fair toaster?

Meanwhile, there has also been mainstream momentum behind a critique of 'free' business models, and their true costs. This has gone from being a niche and quiet concern, to becoming a mainstream conversation between politicians, consumers, companies and judges.

Technology offers us a double-edged sword. On one side, a powerful tool, capable of solving some of our greatest social and environmental challenges; on the other side it seems to invite as many new challenges as it disperses. The trick is in how we wield it.

the waves, is the 'Surface Web': anything findable on search engines such as YouTube or Wikipedia. The rest of it – an estimated 96 per cent – is the 'Deep Web' – anything not findable on search engines, such as medical records. The deepest and final few per cent of this submerged iceberg is the 'dark web' – sites often, but not always, engaging in illicit activity, that can only be accessed with specific software.

Chapter Five
Fossils of the Future

Energy, quantum and parallel realities

I come from a family of Welsh coal miners: my grandfather, Wyndham, was a miner, as was his father (he retired early because of lung disease), my uncle Vivian, and his father. 'Sons followed fathers into coal mining despite the risks, as it was a way of life that they knew with a strong sense of community,' says my mum.

My mother recalls how, when she was three or four years old, her father would come home 'with a black face, covered in coal dust, and he would sit in a tin bath in front of the fire'. He, like most miners, had a very pale face when washed, 'because they never saw the sun'.

Collieries proliferated across south Wales because they had anthracite coal, a particularly good type. Uncle Viv worked the pits, lying on his side in an eighteen-inch gap, 'using a pick, shovel and sledge hammer' to chip away at the coal. My uncle appreciated the 'close camaraderie' with other miners, but says it was physically hard work, and 'in some cases dangerous because the top at times was fragile'.

Coal mining has an ugly history. It was the work given to slaves, and then to leased convicts in a process of 'Slavery by Another Name' that continued well into the twentieth century. Wales was home to the infamous

A Pulitzer Prize-winning book by Douglas Blackmon. Also explored in the documentary *13th* by Ava DuVernay, and the poetic book *Homegoing* by Yaa Gyasi.

miners' strikes through the 1970s and 1980s: a consequence of low pay and dangerous working conditions.
Disasters were common: in 1966, a colliery spoil tip
collapsed onto the Welsh village of Aberfan, engulfing
a school and killing 116 children and 28 adults.

Other fossil fuels – notably gas and oil – also demand considerable sacrifices: disregarding indigenous
ancestral land rights, fracking under people's homes
without their permission, destroying ocean habitats
with oil spills, and increasing conflict. Architect Eyal
Weizman and writer Naomi Klein argue that there is
an 'astounding coincidence' that connects the locations of Western jet and drone strikes with areas of
oil and drought worldwide.

It is now clearly documented that burning fossil
fuels is the leading driver of global warming, causing intensified droughts and erratic weather patterns,
displacing millions of people worldwide and threatening global security.

There is no more urgent need in the world today
than to stop burning fossil fuels.

Our addiction to them is a relatively new thing
For tens of thousands of years, humanity used entirely renewable energy sources. The burning of wood
and other biomass enabled humans to establish settle-
ments; wind and water mills to generate energy from
the air and rivers; and sailing boats to cross oceans;
while sunshine encouraged food growth in green-
houses, and dried wet clothes.

After dipping our toes into the fossil fuel candy
store, getting high and crashing down, we are

increasingly returning to the old ways of producing energy from the elements around us but using
modern technology to access it with greater efficiency and scale.

The work of the Devil?

In 1881 the world's first hydroelectric power station
was built at Niagara Falls. The power generated by
the water's fall of eighty-six feet was used to run local
mills, and light local street lamps. People have long
since tried to mimic the effect, and generate the power
of falling natural water through the construction of
dams and reservoirs, although artificial dams remain
controversial: 25 per cent of rivers now don't run to
the sea, interrupting ecosystems.

New methodologies for harnessing green energy
from water are still being explored: from wave farms
to tidal power, ocean thermal energy and osmotic
energy. Hydropower now provides over 16 per cent
of global electricity – an astonishing 99 per cent of
electricity in Norway. It is claimed that ocean energy
alone, in these various manifestations, could meet
global demand for electricity.

A few years after the Niagara Falls power station
was established, in 1887 Professor James Blyth built
the world's first wind-powered home, in Scotland. He
offered the surplus electricity to the local residents to
light the town's street, but they refused it, calling electric power 'the work of the Devil'. The development

of wind technology since has seen turbines erected in over eighty countries around the world, on hill-tops, valleys and out in oceans, providing 4 per cent of worldwide electric power and growing year on year. As of 2015, Denmark generates 40 per cent of its electric power from the wind.

Meanwhile, geo-thermal energy offers a stable source of renewable energy – it is not weather dependent, and is available 24 hours a day, 365 days a year. It has helped Iceland generate nearly 100 per cent of its electricity from renewable sources, not to mention powered all those beautiful hot springs for which it is famous.

One of the oldest sources of energy, biomass – developed from plants, trees, agricultural and domestic waste – has now become a major player in the move away from fossil fuel energy sources, accounting (in 2012) for two thirds of all renewable energy consumption in the EU. Biomasses are also being used to produce alternative 'green gases' which are carbon neutral and will help to decarbonize conventional gas heating systems – contrary to some political claims that 'fracking' is a necessary transition process for obtaining fuel.

Finally, people are looking directly at our original energy source – the sun. The first solar cells were developed in the 1950s by Bell Labs for spacecraft, but they really started to take off in the 1970s because of growing environmental concern and the oil crisis. The efficiency of solar panels has increased year on year, whilst since 1980 the price has fallen by 99 per cent.

Sunflowers

Fossil fuels, unlike renewable forms of energy
such as wind and solar, are not widely distributed
but highly concentrated in very specific locations,
and those locations have a bad habit of being in
other people's countries.
Naomi Klein, *London Review of Books*,
'Let them drown'

In front of me as I type, at the off-grid house I have
rented in the mountains below San Francisco, are two
large silver solar panels, arching their necks up towards the sky like sunflowers. Solar panels can now
be seen all around the world – from the rooftops of
sci-fi buildings like Apple Park, to parking meters in
London, or alongside simple mud huts in Africa. Solar
panels' flexibility and increasing affordability gives
them a leading role in another element of the energy
revolution: decentralization.

Historically, consumers had limited (if any)
choices for their energy supply, being dependent on a
grid with vast and expensive infrastructure – mines,
oilfields, pipes. Yet now, new decentralized technologies like solar are increasingly empowering people to
make choices around personal energy use: by switching energy supplier where possible or even installing
their own solar panels or geo-thermal units, using
biomass, or investing in local community energy
cooperatives.

'One of the great things right now about clean energy is that it's actually cheaper than dirty energy,' says Apple's Lisa Jackson. 'You're starting to see communities, even politically conservative communities, who say, Look, I want to be able to choose the energy I want, I shouldn't have to buy from a monopoly like a utility, who's going to sell power that's more expensive than the clean energy that makes so much more sense.'

In the United States, multiple companies are making the renewable transition for consumers easier through programmes with no upfront installation costs. For example, 'Dandelion Energy' – an offshoot of Google X – builds home geo-thermal units in New York through a leasing programme, and SolarCity have pioneered solar panel leasing plans across the country.

My daughter's grandparents in south Portugal live in one of many households who have installed their own solar panels to meet their energy needs, selling the surplus back to the grid. Meanwhile, my late uncle Adam Giffard in Devon was a pioneer in developing woodchip pellets, as a renewable fuel source in the UK. As long as trees or plants used for biomass are replaced, they offer a local, carbon-neutral supply of energy.

From Bristol to Brooklyn, cooperatives have been set up to pool local resources into renewable installation. A report by the UK's Department of Energy and Climate Change claims that community-owned renewable energy projects bring twelve to thirteen times more value back to the community compared to private schemes. In Germany in recent years, nearly half of renewable energy installed was by citizens, and according to a report commissioned by Greenpeace, by 2050 almost half of EU households will be involved in producing renewable energy.

Decentralization is shaping the energy landscape in lower-income countries. Travelling around Mozambique, I was struck by the number of solar panels we saw. It makes little sense for emerging economies to invest in large infrastructure, digging up land and building complex grids to support their energy needs, when solar offers a cheaper and better alternative. Communities that have so little – a handmade mud hut, a few simple pieces of clothing, some chickens – are leapfrogging generations of energy technology to light their homes with solar – and perhaps also, bypassing landlines, to charge their mobile phones.

Storing the sun

We can get our energy from the Sahara if we
actually solve the right problems.
David Attenborough

The intermittent nature of renewable energy supplies – you only get solar power when it's sunny, and wind power when it's windy – has presented a significant barrier to growth, but this is where innovations in batteries come in.

The first rechargeable battery was invented in 1859 by a French physicist, made of lead-acid. In 1900 a nickel-cadmium battery was invented, and these two types dominated the market through the twentieth century. More recent developments, spurred on by space technology, such as lithium-ion, are finally making solar and wind into viable mainstream energy solutions.

Battery giga-factories are being built across Europe, and many state-level national grids are investing significantly in their storage capacities.

In 2016 the small Samoan island of Ta'ū, where Margaret Mead first made her name as an anthropologist studying adolescent girls in the 1920s, switched from relying on 11,000 gallons of imported diesel each year to a 100 per cent solar energy supply. Having installed 5,328 solar panels and 60 Tesla Powerpack batteries, the island can now store enough energy to last three days without sun.

Meanwhile, concerned about the finite materials used in the lithium-ion battery, engineers and scientists have been developing alternatives. Researchers at the UK's University of Surrey claim to have developed a polymer alternative to lithium which will make batteries 1,000 to 10,000 times more powerful. Google X helped develop the ideas of Nobel Prize-winning physicist Robert Laughlin to store energy using molten salt. (Yes, ordinary salt!)

In 2014, Harvard professor of chemistry Alán Aspuru-Guzik was part of a team with Roy Gordon and Michael Aziz who developed 'Flow' batteries, using a variant of a common molecule to store energy. They are currently working on a third-generation model of the Flow battery, which is getting close to being commercialized. According to Aspuru-Guzik, advances in the capacities of energy storage will offer the 'next revolution' in green technology.

Can quantum computing robots solve climate change?

Ask Aspuru-Guzik what he thinks will save the planet and he will give you a surprising answer: 'Robotization of the laboratory.' Aspuru-Guzik is working on developing a robot-run laboratory. Their first robot, nicknamed 'Bob the Bot', is mixing margaritas because, Aspuru-Guzik jests, it needs to get used to handling liquids.

Human scientists can only move so quickly in terms of making discoveries – they need to eat, sleep, check their emails. Instead, robots – empowered by AI and quantum computing – will be able to innovate at a wildly accelerated pace, creating 'exponential growth in science and technology'.

What does exponential growth look like, and what might that mean for climate change? Hartmut Neven, the founder of Google's Quantum Artificial

Intelligence Lab, and his team have achieved not just exponential growth but double exponential growth, since named 'Neven's law'. When I visited him in the Los Angeles office, he was buzzing with the excitement of a child.

They had just made a breakthrough, he told me in hushed German tones. Hartmut recently had to call Google's systems administrators to let them know that in order to check the computations of their new quantum chip, a classical computer would require the electricity of a whole city. And its power was doubling every fortnight.

A few months later, in October 2019, Google announced in *Nature* that this work had enabled them to achieve 'quantum supremacy': their processor was able to compute in 200 seconds what would take a supercomputer 10,000 years.

> Understand . . . well as I may, my comprehension
> can only be an infinitesimal fraction of all I want
> to understand.
>
> Ada Lovelace

Since their invention by Ada Lovelace and Charles Babbage in the nineteenth century, computers have run on the logic of dualistic possibilities (1 and 0), reflective of the classical Newtonian world view. Yet Lovelace was the first to recognize that computers had the capacity to move beyond pure calculation, and since then the classical Newtonian understanding of reality has evolved.

The physicist Richard Feynman argued, 'Nature isn't classical, dammit, and if you want to make a simulation of nature, you'd better make it quantum mechanical, and by golly it's a wonderful problem, because it doesn't look so easy.'

'The most straightforward interpretation of quantum physics is that we live in a multiverse, made up of many parallel universes,' says Hartmut, casually. Quantum computers are 'the first technology that takes seriously the notion that our world is structured as a multiverse . . . the quantum processor essentially uses a number of parallel universes to do its computation'.

Sound confusing? Put simply, the quantum computer is able to consider and evaluate many, many possibilities, if not all possibilities arising within a system. This creates a paradigm leap for invention and problem solving. 'Could quantum computers help with one of the defining challenges of our generation – climate change?' he asks. 'We think the answer is yes.'

'We are going towards 8 billion people – if all those people want anything near a Western lifestyle, we don't have the industrial foundations to do that without turning the Earth into a desert. So, we need to re-engineer our industrial foundations quickly – and a very useful tool (not the only one) for doing this is a quantum computer because it will allow us to create the new materials we need, literally from the atoms up, for a more ecologically sustainable industry.'

When you investigate the world at the ridiculously small scale of atoms and subatomic particles, things can get weird. Take the famous Schrödinger's cat thought experiment, used by Nobel Prize-winner Erwin Schrödinger to demonstrate that particles can exist in a combination of multiple states at the same time. A cat is locked in a box with some radioactive material. If the material randomly decays, the radiation triggers the release of a poisonous gas, killing the cat. Until one actually looks inside the box, the cat exists in two states simultaneously: alive and dead.

For example, right now, if an engineer has an idea for a new electrolyte material to improve batteries, she has to build the battery, then take it to a laboratory to test and measure. Because of the high costs in time and money of this process, only a few hundred electrolyte materials get tested every year, but 'if you had a quantum computer you could simulate the process – you could check millions of battery materials and take only the most promising ones to the lab'.

'Many engineering disciplines – whether it's batteries, solar cells, super-conductors, catalysts in pharmaceutical industry, ecologically friendly plastics, or reducing the amount of energy needed for fertilizer production: there's a long list of areas in industry where a quantum computer would help us remake the industrial foundations of our society quickly – hopefully quickly enough.'

Quickly enough is the key issue. How quickly can we move?

Moon-shots

A team of researchers in Stanford estimated that by 2031 to 2051 the whole world could be powered by renewable energy, focusing on solar and wind and using existing technology, if there was concerted political will to do so. The costs would be comparable to those of conventional energy, but the transition would require a 'large scale transformation – comparable to the Apollo Moon project or constructing the [US] interstate highway system'.

An estimated $100 billion was spent on the Apollo space programme. Compare this to the fact that since the 1970s, the US government has invested around $3 billion in research and development of solar panels. The challenge is in many ways a question of political priorities. The environment needs to be treated as a public good, akin to healthcare and education.

We will discuss the politics around the transition in Chapter Nine, but if we want to look for political inspiration that this transition is possible, we can turn to Costa Rica or Iceland who source 98 to 100 per cent of their electricity through renewables. Many countries and states around the world have made commitments to generate 100 per cent renewable energy in the near future, including California, Hawaii, and many cities in the US and UK.

Companies are at the vanguard of the drive towards clean energy – crossing geopolitical borders and disparate political commitments. At least one hundred and thirty of the world's largest companies (from General Motors to Nike) have publicly committed to using 100 per cent renewable energy by specified dates – as part of an initiative called RE100. Considering that industry accounts for two thirds of end-of-use electricity, this movement is having a formidable impact.

Illina Frankiv, Global Energy Program Manager at the largest co-working company in the world, which is part of RE100, tells me that it has been challenging to try to get the company to source

100 per cent renewable energy worldwide. It often requires installing their own solar panels, or building offsite solar/wind, when utilities don't offer 100 per cent renewable energy off the shelf. Yet she is very optimistic about the trend of corporations getting into the business of energy generation to serve their own needs: 'This large demand and action will reduce the cost of renewable generation, making the transition for utility companies inevitable.'

'All companies should strive to become carbon positive: their efforts should result in greater carbon offset than emissions,' says Illina. Alongside pushing towards 100 per cent renewable energy, her company offset all of their corporate flights via forestry projects. They are also looking to improve the energy efficiency of their buildings, and refuse to purchase meat for any employees: 'Some think it's a draconian rule, but it has a great carbon avoidance.'

Ambitious corporate initiatives also offer consumers the chance to affect the global energy landscape, by supporting companies heralding positive change in countries where consumers would otherwise have no political vote.

Take China as an example. As a Western environmentalist it is hard to know how to have a voice in China's energy policies. Yet, being mindful of the companies we buy into – supporting ones which are working on carbon-neutral supply chains in China and elsewhere – offers us that possibility.

Investment into renewable energy now exceeds investment into non-renewable energy. Meanwhile, campaigns around the world – from college campuses to the Rockefeller Foundation, or World Council of Churches – have seen trillions of dollars divested from fossil fuels, making it the fastest growing <u>divestment</u> movement in history. Other activist investors take a different approach: becoming shareholders of fossil fuel companies so they can try to direct the companies towards cleaner energy from the inside.

Dale Vince – electric-motorbike-riding vegan and founder of leading renewable energy company Ecotricity – is optimistic, pointing out that green energy has gone from being something that 'a few hippies and activists were interested in' in the 1990s, to a mainstream industry. The UK has moved from 2 per cent to 30 per cent renewable energy and China is 'building a windmill every minute'. 'It's stupid to write off the future based on what is possible in the present,' Dale says. 'I am always optimistic because we are an inventive species.'

In 2020, the International Energy Agency announced that energy-related carbon dioxide emissions plateaued from 2018 to 2019, because of a decrease in coal, and increase in renewables. 'We now need to work hard to make sure that 2019 is remembered as a definitive peak in global emissions, not just another pause,' said the IEA's executive director.

The holy grail: faking the sun

The apartment I lived in during my twenties, in London, was connected to a train station. I once tried to persuade my building to switch to a renewable energy supply but found that, short of moving house, we had no choice but to use the train station's supply of French nuclear energy to power my fridge, lights, heaters and even charge my electric car.

How did I feel about this? Well, better than if I was being forced to use coal or oil, knowing that the carbon emissions from nuclear are much lower. But it always felt like a problematic solution.

Nuclear power sums up humanity's shifting love affair with technology. Nuclear is arguably the most controversial form of energy creation to have emerged in the twentieth century: the clean dream of the future and the nightmarish scourge unleashed by Chernobyl or Fukushima. The world is polarized: whilst Germany is divesting from nuclear, new plants are springing up elsewhere.

Many of us have a visceral aversion to this form of energy as a result partly of ongoing anxiety about nuclear weapons, and also in the wake of such disasters. Yet nuclear power plants remain statistically safer to work in than coal mines or oilfields. Accidents are rare, though of course when they do happen, the results can be catastrophic.

It is often touted as the most viable replacement for fossil fuels in generating electricity without incurring high carbon emissions. Indeed, many politicians and scientists have followed this logic, and it has been the most scaled energy since the Second World War, now generating 10 per cent of the world's electricity. Even some environmentalists have been convinced of nuclear as the only viable large-scale way to move away from fossil fuels as quickly as we need to.

In 2013 four eminent scientists from MIT and Columbia University wrote an open letter calling on the environmental community to accept the role that nuclear needed to play in addressing climate change: 'While it may be theoretically possible to stabilize the climate without nuclear power, in the real world there is no credible path to climate stabilization that does not include a substantial role for nuclear power.'

Giving some scientific backing to the Buddhist dictum that everything is interconnected, in the 1970s, chemist James Lovelock and biologist Lynn Margulis presented the impactful theory that Earth is a single self-regulating system, whereby life manages the planet for the survival of life.

James Lovelock, originator of the Gaia theory, has also been a surprising voice in the pro-nuclear lobby since 2004. His justification is the urgency of the crisis: 'We have no time to experiment with visionary energy sources; civilization is in imminent danger and has to use nuclear – the one safe, available, energy source – now or suffer the pain soon to be inflicted by our outraged planet.'

Yet there are big issues with nuclear which I find too problematic to overlook. Nuclear energy is generated by splitting uranium atoms, which means that nuclear fission is *not* a renewable energy given that uranium is a finite resource (estimated to run out in 135 years).

Nearly a third of the world's uranium deposits are in Australia, and the indigenous communities who lived near them always called them the 'sick land' – long before the uranium was discovered.

Jeffrey Lee – Djok clan leader and sole custodian of an area known as Koongarra – and his ancestors resisted years of pressure from a French nuclear company who wanted to mine the uranium there. Instead of capitalizing on this partnership – worth billions of dollars – he donated the land to form part of the UNESCO World Heritage Kakadu national park where he works as a ranger.

'There's poison in the ground. It's my belief that if you disturb that land bad things will happen,' Lee said. 'I have said no to uranium mining at Koongarra because I believe that the land and my cultural beliefs are more important than mining and money. Money comes and goes, but the land is always here.'

Researchers around the world are working on a 'fourth generation' version of nuclear fission energy that wouldn't depend on virgin uranium. Ongoing research into breeder reactors, for example, offers the possibility of recycling existing deposits of uranium, and using uranium taken from decommissioned weapons. Yet nuclear's afterlife remains an issue: radioactive waste is often flown around the world, to be buried in the oceans of lower-income countries – additional sacrifice zones – labelled with an icon that looks a bit like a scorpion, to warn whoever finds it, hundreds of years from now, against opening the containers.

Nuclear fusion, in which two or more nuclei are fused together under conditions of immense heat and pressure, has long been held as the holy grail of energy production since it offers to tap into a non-carbon emitting and plentiful energy source, that is much safer than fission. Like fission, fusion requires less land than solar or wind, and is not weather dependent, though it is hard to envisage it facilitating the decentralization of energy ownership.

Nuclear fusion mimics processes that occur in the sun, our original energy source. It depends primarily on hydrogen, a molecule that we have in almost limit-less supply, enough to last several million years. This also produces radioactive waste but, unlike that of fission, it is quick to decay.

After decades of hope, the nuclear fusion industry has renewed optimism as investment pours in from both the private and public sectors, and companies claim they will be operational by 2030–40. In 2019, the British government invested £220 million into their own nuclear fusion power station, and ITER (meaning 'the way' in Latin) is an ambitious international nuclear fusion engineering project in the South of France that has been funded by the EU, India, Japan, China, Russia, South Korea and the US.

Nuclear fusion is '*the* climate solution', Moritz von der Linden, CEO of Munich-based Marvel Fusions, tells me emphatically when we meet. Moritz doesn't believe renewables alone will get us as far as we need to go: 'Everything we are doing right now is not sufficient to meet Paris climate goals. Fusion is for me the perfect energy source – nothing else is even close.'

> You have to backstop [renewables] with
> something, and that something, in my view,
> should also be clean – it should be nuclear –
> and then let's let the battle rage: for which is
> the cake and which is the icing.
> Andrew McAfee, *More from Less*

Nu-clear remains un-clear for me. We arguably do need it right now to decarbonize our economies quickly, but let's not lose sight of the fact that virgin uranium is finite. I welcome R&D that seeks to make nuclear a viable, scalable, long-term solution. That said, I was very happy when I moved to a home where I could buy electricity powered by the wind and sun, and I hope they're not just icing.

Saving us

> Over time all energy generation will either be renewable, or we'll have economic collapse. Civilization will power down, essentially.
>
> Elon Musk

Short of turning off all power and embracing the dissolution of civilization as a good thing (which some hard-core environmentalists might hope for), it is clear that we must urgently find renewable ways to meet our energy needs. We know this is technically possible, and indeed progressing, but don't let my optimism blind you to the reality of how much work there still is to be done: an unnervingly strong fossil fuel legacy and lobbying to fight, entrenched interests, existing infrastructure, sunken costs, and lower-income nations who want (justly) to prioritize their economic growth, whatever the environmental cost. Hence why we continue to burn absurdly high levels of fossil fuels, and warm our frog bath, despite all the scientific and political awareness that we shouldn't.

Yet I remain optimistic because the rise of new technology, grass-roots decentralized power generation, corporate renewable ambition and a growing political mandate, suggest that the burning of fossil fuels is a dying game.

Participating, as individuals, in the global energy matrix is essential but intensely complex as our energy touchpoints are wide and sometimes insidious. It is not as simple as buying sustainable fashion or decreasing the amount of meat we eat. The most obvious thing is to change home energy supplier and speak to any institutions (companies, colleges, NGOs) about doing the same.

Since the Paris Agreement on Climate Change was adopted in 2016, UK banks and pension funds have invested nearly £150 billion into financing fossil fuels (not to mention investments into the arms trade), but there are ethical alternatives, such as Triodos bank, and polls show the majority of people in the UK want their investments to be responsible and fossil-fuel free.

Then it is important to consider how much energy our lifestyles actually require, and how we can decrease, and offset, our energy needs. How about changing banks and pension providers to ones that do not invest in fossil fuels? (Or the arms trade ideally.)

Finally, and perhaps most importantly, is the political piece of the puzzle. Can we afford to go green? Can we afford not to?

Worldwide, fossil fuels are being directly subsidized at nearly four times the rate of renewable alternatives. The International Institute for Sustainable Development (IISD) argue that redirecting 10 to 30 per cent of the fossil fuel subsidies would be sufficient to pay for a global transition to clean energy. 'Almost everywhere, renewables are so close to being competitive that [a 10 to 30 per cent subsidy swap] tips the balance, and turns them from a technology that is slowly growing to one that is instantly the most viable. It goes from being marginal to an absolute no-brainer,' says Richard Bridle of the IISD.

The battle against fossil fuels is not yet won, but the tide is turning, like the Moon that pulls it. The transition is technically possible. What we are missing is enough political will: to act as a counter-force to lobbying. Yet in democracies this is formed by us, the people. If there is enough political will to fly us to the Moon – which incidentally captured the photo of planet Earth, a small fragile oasis in vast space, that inspired the mainstream environmental movement – we can create enough political will to bring us back down to Earth.

Chapter Six
A Spacefaring Species

Travel and the paradox of adventure

I felt a sharp, hot slap across my inner thigh. *Don't panic.* I swam back to the shore, as fiercely as my leg would allow. On the beach, I inspected the wound, several inches wide, like a handprint from *The Blair Witch Project*. My friends were nowhere to be seen. I tried to walk back to our house, but my head hurt and reality morphed with the waves of heat and nausea. *Must find humans.*

A ragtag group of local residents and tourists came into sight and I threw up on the sand. A woman leaned over: 'Portuguese man-of-war,' she said. It was aptly named, considering this land was colonized by the Portuguese a few centuries ago and I was feeling cross with my Portuguese boyfriend that day.

Speaking with the locals later, we learned that my attacker, a poisonous siphonophore similar to a jellyfish, had proliferated in the area over the previous year. Quite a few things had changed since our last visit three years earlier. We were now being eaten alive by abundant swarms of mosquitoes, and the fish had almost disappeared. The rising sea had swallowed up a beach bar we used to visit, and left a long line of white and blue plastic glittering in its wake.

Perhaps the only thing that hadn't changed in this corner of Brazil was the presence of tourists, like us, travelling the world in search of Good Times.

Good time for the end time

We are instinctively nomadic. Adventure and curiosity have underscored humanity's journey from the savannahs of Africa, to the tips of the Arctic poles, through to our burgeoning desire to explore new solar systems. Movement would appear to be in our DNA – a biological function of evolution, as microbes and species compete to colonize new corners of the natural world.

Our feet, boats, horses and the wheel were the mainstay of travel until the discovery of large supplies of coal began to revolutionize and accelerate the way we move. It is perhaps ironic that the contemporary bastion of eco-friendly travel – sailing boats and trains – were actually respective catalysts of Colonialism and the Industrial Revolution: strange how the meanings of symbols morph.

As modes of transport have become faster and more comfortable, our instinct to travel has fuelled us, and all of our billions of tons of stuff, to draw circles around the planet. Travel has enabled us to cross-pollinate cultures, expand world views, and share knowledge, customs and inventions. I was brought up listening to stories from my mother about hitch-hiking across continents, and of how my father built his own boat and sailed the world. I grew up itching to travel further than our family camping trips allowed, and feel deeply grateful for the chances I have since had to visit so much of the world. I still long to see more.

Yet increasing awareness of the impact of travel on our deteriorating environment has provoked many clarion calls for reduction: *drive less* and *fly less* being mainstays of the environmental conversation. Whole new adjectives have been invented to capture this cultural shift, such as *flygskam*, *smygflyga* and *tagskryt* – Swedish for flight shame, flying in secret and train bragging.

Should we stop moving so often? Or should we move differently?

The art of moving slowly

Only ideas gained from walking have any worth.
Friedrich Nietzsche, *Twilight of the Idols*

When John Francis watched 850,000 gallons of thick black oil pour into the San Francisco Bay in 1971, after two tankers crashed, he made a pledge not to travel by car, plane, bus or train again. He wanted to take some responsibility for the oil that was washing up on the shore. He took up the oldest mode of human transport, and a favourite of many philosophers: walking.

At first, he thought he was going to start a movement, but the dawning realization that 'I was alone!' didn't stop him, and he continued to walk everywhere for twenty-two years. Even the arguments he had with his friends and neighbours, that one person couldn't make a difference in such a huge problem, didn't dissuade him. He spent seven years crossing the United

States by foot and attended university, earning a PhD in land resources.

I met John for a cup of tea near San Francisco, where he told me, 'People would drive by me and say, John, I feel so sorry for you – let me give you a ride, and I'm like, don't feel sorry for me, I feel really great that I've stepped out of the automotive track.'

His journey has inspired others to re-frame travel in different ways: 'Recently someone wrote me and said, "I was just four years old and I remember you were walking and playing the banjo . . . now I'm twenty-six years old and I don't ride in cars and I ride a bike – I was inspired by you." I'm honoured to be remembered that way.'

Nietzsche, Thoreau, Kant, Rousseau, Rimbaud and the Brontë sisters all wrote of the joys and benefits of walking. Surrealist, Fluxus and Situationist artists advocated walking as an artistic and even political practice: to foster chance, serendipity and an appreciation of random everyday moments. The spiritual tradition of pilgrimage by foot can also be found in many countries and cultures.

> I'll walk where my own nature would be leading.
> Emily Brontë, 'Stanzas'

I love walking, and encouraging more of it can only be a good thing: for personal as well as planetary health. But thinking pragmatically about travel, the reality is that many people continue to want – or even need – to move quickly. Very quickly.

When I met the San in Botswana, I distinctly remember one man saying, 'I want a Toyota,' and the Yawanawá tribe in the Amazon told me how the diesel-engined boat transformed the passage from their village to the next town: 'What we did in five days, the engine took me in six hours,' Chief Nixi-waka told me. 'That's a gift. I have to be thankful for this invention.'

There is no doubt that the inventions around travel have fulfilled a universal thirst for speed. The question is, can we make our inventions more sustainable?

The business of moving fast

Mention technology and sustainability in the twenty-first century, and invariably one name will eventually come up: Elon Musk. It rings more like a sci-fi perfume brand than a person, and indeed, he is often treated as non-human. *Elon Musk will save the day*, goes a common tech-assumption, as if he were an alien, an avatar or Artificial Intelligence itself. The South African engineer's vision, and his role spearheading green technology from Tesla to SolarCity, SpaceX to Hyperloop, alongside his apparent appetite for generating controversial headlines, have made him a polarizing figure.

'Elon Musk is not a person but a species survival strategy,' writes journalist David Wallace-Wells. Another article, 'The Church of Elon Musk', describes his rapturous reception at a tech conference. Elon embodies perhaps the broader – and equally polar-izing – Church that humanity is increasingly looking to for salvation: technology.

Will it save us or destroy us? Elon threads both these narratives, as optimistic about sustainable energy and space travel as he is concerned about AI's ability to render us obsolete.

One misty day in San Francisco I went to Elon's (large, apparently haunted) house to chat 'Mars and cars' with him.

An electric Trojan horse

'The whole point of Tesla is environmental,' Elon told me. 'When we started, the goal was to get the industry to move towards electric cars. There was literally nothing – not even short-range cars. Electric vehicles had been written off.'

Hands down, Tesla have changed the car landscape. They have achieved the goal of making a long-range electric car both viable and desirable, and have encouraged a wider shift across the market with most automotive companies now making or planning electric cars. They are also developing an electric pick-up truck and semi-truck for road cargo transportation.

Their work is critical considering that cars are the biggest contributor in the wider travel-carbon matrix which is responsible for more than half of global oil demand and contributes approximately a quarter of global carbon dioxide emissions. In the UK, travel has overtaken energy as the number one contributor to carbon dioxide emissions.

A revolution in infrastructure – namely, the way we travel and transport goods – is also an essential piece in the puzzle to achieve 100 per cent sustainable energy. Whilst we are still pulling the trigger on oil pumps to fuel our cars, trucks, ships and planes, we will not absolve our addiction to fossil fuels.

Holding down that trigger to physically pump petrol into my old 1960s car had always felt like my most visceral, and clearly guilty, relationship to fossil fuels. I used to have dreams about it, my subconscious nagging at me.

'They would never allow petrol cars today,' Elon mused. 'If someone said, we are going to give you this toxic, dangerous, flammable chemical that you have to pour into a tank and go to these special, chemical refuelling stations, and you'd better not smoke when you're filling it up because it will catch on fire . . . And then if you crash, the liquid spills on the ground and catches on fire. That sounds crazy.'

Bio-fuels were once looked to as a way to 'green' cars, but the consequent land-rush to grow palm oil and soya oil has driven high levels of deforestation around the world, and their potential has been reappraised. 'About half the palm oil coming into the EU is going into the tanks of people's cars,' John Sauven, executive director of Greenpeace UK, tells me bleakly, reminding me that the path to hell is often paved with good intentions.

So, short of everyone ditching their cars, electric vehicles offer a promising solution to upgrade our travel and interconnected energy systems. An EU study found that an electric car using electricity from a power station fired by fossil fuels would still use two thirds *less* energy than a normal petrol car, and as our grid systems move in the direction of renewables, many electric cars could soon be run 100 per cent renewably.

Is Tesla's electric car a Trojan horse, holding within it a wider drive towards renewable energy? Tesla have long claimed they are not a car company but a sustainable energy company. In 'The Secret Tesla Motors Master Plan' published online in 2006, Elon wrote, 'The overarching purpose of Tesla Motors is to help expedite the move from a mine-and-burn hydrocarbon economy towards a solar electric economy, which I believe to be the primary, but not exclusive, sustainable solution.' This was staged via a three-step pricing plan:

'The master plan is: Build sports car; Use that money to build an affordable car; Use that money to build an even more affordable car; While doing above, also provide zero emission electric power generation options. Don't tell anyone.'

'To what degree has Tesla accelerated the advent of sustainable energy?' Elon asked when we met. 'Let's say, but for Tesla, it would have taken ten years longer. That's how I would measure the good of Tesla.'

Electric cars are not the perfect environmental travel solution (walking, bicycling, sailing or simply *not travelling* are superior), nor are they without eco-criticism. 'What is the best way of wrecking a city? Pour cars into it,' George Monbiot argues in his critique of 'carmageddon': cars clog up city streets, require land for roads, increase traffic, and take valuable city space that could otherwise be used for communities to thrive in. Different cities around the world – Paris, Barcelona, Seattle, London – are considering or implementing car-free zones or car-free days to create community-rich spaces, and save hundreds of lives from air pollution.

Tesla have been fined multiple times by the EPA for hazardous waste and air pollution and people question the materials inside batteries, though Tesla claim that theirs contain no hazardous compounds, and Elon says, 'You can recycle essentially the whole battery.'

Although electric cars produce no emissions when driving, the *production* of an electric car typically requires more energy than a normal car. Studies suggest that bearing this in mind, over its life cycle an electric car (charged with non-renewable electricity) will be responsible for an estimated 80 per cent of the emissions of a petrol one – which makes it better than buying a non-electric new car, but not necessarily better than keeping an existing one running. Thus, some environmentalists claim that the best type of car is your old one.

It was with these questions in mind that I spent a long time trying to decide whether to sell my car, and buy an electric one instead. I test-drove every electric car on the market in 2015, and Tesla was the only one offering long-range at that time. It stretched my budget but I decided to buy it, as I believe that if you have the luxury of choice, it is important to invest in the infrastructural shift away from petrol stations, to help facilitate the wider future transition towards sustainable energy. As per Tesla's Master Plan, as demand increases, this choice will no longer be a luxury one. Indeed, when they released their lower-priced Model 3, it hit the top ten best-selling car lists in the UK and the US.

Driving a Tesla, you are driving a robot. With self-autonomous driving features, you can cruise down motorways, foot off the pedal, the wheel steering itself, with the car in control. The car also parks itself: a joy for a notoriously bad parker like me.

Multiple companies (like Google's Waymo) are now developing fully autonomous cars, and driving licences will likely be irrelevant by the time my daughter is grown up. There is a subtle environmental promise in this transition, machines being more energy efficient as drivers than humans.

Tesla's innovative software is not without bugs – and risks – earning them the common nickname Stressla. In spite of this, I still love the car and we have travelled across, up and down the UK together.

Yet I often wonder, will my Tesla still be as elegant and functional half a century from now, as my 1969 car has been? Tesla aren't clear about how long their cars will last, but like most manufacturers their business model is haunted with the potential for planned obsolescence. When I asked Elon this, his answer was

Although a green technology boom is arguably welcome news, we have to push for fairer trade behind it. The world's largest technology companies have been named in a legal case brought by Congolese families who say their children were maimed or killed whilst working in cobalt mines.

Meanwhile, it is estimated that 1.5 trillion litres of water will be lost in Chile's Atacama salt flats in coming years, through lithium mining: 'My ancestors made a spiritual pact with their territory,' says Sonya Ramos, a local indigenous leader who has won the support of an environmental court in her community's protest against the mine.

optimistic if complicated: 'You'll probably have to recycle the battery pack every twelve to fifteen years. But then the battery pack you get in twelve years will be way better than the one that's in there today.'

There are concerns that finite materials in batteries – cobalt and lithium in particular – impose a hard limit on the potential for expanding electric batteries and therefore renewable energy storage. One report claims that if all the drivers in the world converted to electric cars there would not be enough lithium to meet global demand for more than seventeen years. That said, as we saw in the last chapter, battery innovations are progressing rapidly – in large part driven by Tesla – and Elon maintains that lithium is in no short supply. 'You don't even need a supernova to create lithium . . . piece of cake!'

Hartmut Neven thinks innovations for batteries are a good place to start applying quantum simulation techniques because existing battery technology is not sufficient to meet the scale needed, due to limits on cobalt supplies: 'When I hear governments passing laws like, By 2025 all cars should be electric, I feel like, Are you sure? Have you done the basic numbers? There is nobody who will give you the number of electric cars you will need.'

Meanwhile, 'green hydrogen' is being championed by others – namely, Japan and Shanghai, China. Hydrogen fuel cell technology converts liquid hydrogen into electricity, producing only water as a by-product, and no carbon emissions. They can be refuelled in five minutes, although the lack of hydrogen refuelling stations presents a barrier to growth.

Following on from the success of the electric hybrid car, the 'father of the Prius', Toyota chairman Takeshi Uchiyamada, has been at the helm of hydrogen-fuelled cars. He believes that by 2050 nearly a third of vehicles will be powered by hydrogen. Meanwhile, in 2018 Germany introduced its first hydrogen train, with more planned across Europe.

Electric vehicles represent only 2 per cent of the cars currently being sold, with hydrogen significantly less, but this trend looks set to accelerate as the technology advances, prices drop and political policies incentivize – or even force – change. Many subsidies and benefits for low-emission vehicles exist worldwide, and many countries from China and India to large swathes of the EU have announced forthcoming bans on the sale of cars powered by fossil fuels.

'Vision is easy, execution is hard. The idea of getting to the Moon is easy. But getting to the Moon is hard,' Elon muses. 'The idea of powering an electric car existed for a long time, 150 years in fact. The issue was not the idea. The difficulty was making an electric car company succeed and not go bankrupt. So far, so good. Pretty dicey!'

Electric world

Our world is going electric: trams, scooters, bikes and skateboards are springing up on city streets. Three quarters of railway lines around the world are already electric, and buses are going that way too. The Chinese megacity Shenzhen has established a fleet of 16,000 electric buses, consequently nearly halving carbon dioxide emissions and fuel bills. More than thirty Chinese cities aim to achieve 100 per cent electric transport by 2020; London aims to make all its buses either electric or hybrid by 2020; and New York all electric by 2040.

Electric cars have put a wire in the water and generated ripple effects for boats too: essential as boats have a huge – if often unseen – impact, and lie in the murky waters of international law. A cruise line is developing a hybrid engine, a hopeful response to the galling fact that cruise ships generate three to four times more emissions than jet planes (consider watching Hasan Minhaj's insightful critique of <u>cruise ships</u> before you jump on one).

'Cruises, the answer to the age-old question: What if porta potties had swimming pools?' so begins comedian Hasan Minhaj's near-half hour skewering of cruise ships and their awful track records for water and air pollution, worker abuse, guest safety, punishing crimes committed on board and tax dodging.

Then there is the question of how our 'stuff' moves around the world. Ships carry 90 per cent of world trade, fuelled mostly by crude-oil tankers, causing 2 to 3 per cent of global emissions. Electric cargo ships have been developed in China and Norway. In 2018 electric, emission-free container barges began to operate from Amsterdam and Belgium, using inland waterways. It is expected that they will remove the need for 23,000 trucks – usually run on diesel – as a result. The UN shipping agency has agreed the industry must cut shipping emissions 50 per cent by 2050, compared to 2008 levels.

In Ireland, I once met Jorne Langelaan, one of the founders of another Dutch company leading the way in emission-free sea travel: Fairtransport have been delivering cargo across the Atlantic by sailing ships for ten years; and Jorne's new project, EcoClipper, aims to sail both the Atlantic and Pacific 'to finally connect all continents with a global [emission free] shipping line'. He and his girlfriend Elise had travelled from Amsterdam to Ireland by ferry and land, as they refuse to fly.

Icarus

You haven't seen a tree until you've seen its
shadow from the sky.
Amelia Earhart

In my favourite dreams, like a bird, I fly.

When my mum was young, she longed to fly. She once sewed together chicken feed sacks, holding them aloft like a small parachute, as she jumped from a bank. She had never seen a plane.

In Faith Ringgold's *Tar Beach*, an eight-year-old girl in 1930s segregated America lies on the tarmac roof and imagines herself flying over the city.

Flying began as a utopian dream. Long before Leonardo da Vinci's sketches of helicopters, balloon-omania, or Amelia Earhart's solo flight across the Atlantic, humans have imagined taking to the sky.

An eighteenth- and nineteenth-century fad in hot-air balloons attracted daredevil performers and gathered huge crowds in France and England. It was particularly popular with women, such as 'Lily' Cove ('Leaping Lily') who became known for pulling off her skirt to reveal bloomers mid-flight, but who tragically passed away, aged twenty, when her parachute failed to open.

Nothing is riddled with paradox and polarization so much as our desire to fly. Flying seems to bridge humanity and the heavens. It has knitted up our world, and now looks set to see us living in space.

Are planes glitter in the sky that da Vinci would have wet himself to see? Or are they symbols of decadence and impatience, streaking invisible pollution that we breathe? Will our wings get burnt? Will we fly into our own extinction? Or might flying actually, weirdly, help us to survive?

When Greta Thunberg was born on 3 January 2003, I had just turned fifteen, and was soon to tell my school I was sick so I could fly to New York and LA. I had started modelling the year before, and my jet-setting international career quickly followed. I took any opportunity I could to travel to remote and exotic places, always trying to stay a little longer to absorb the culture.

It was a mind-expanding experience that has shaped me as a person. I feel I learned more by being exposed to different communities and cultures than I did in school books. Listening to Pink Floyd, watching the planet move beneath the clouds, offered sublime moments. But as my environmental awareness has grown, so has my guilt: the last plane I took, I watched the clouds whilst listening to David Wallace-Wells's *The Uninhabitable Earth* and the journey felt more racked with shame than sublime.

In 2015, when Greta was twelve and had made the personal decision to stop flying because of climate change, I had already travelled in most continents.

A Spacefaring Species

That summer I gave birth to my daughter. Like many people, I have significantly reduced my flying, and offset when I do, but have often wrestled with the question: should we stop flying altogether? I find it highlights my contradictions to the core.

> If you regularly fly around the world, eat meat and dairy, and are living a high-carbon lifestyle – that means you have used up countless people's remaining carbon budgets – carbon budgets that they will need in their everyday life for generations to come.
> Greta Thunberg, June 2019, speech at the Brilliant Minds conference in Stockholm

Greta's words pierced the room of entrepreneurs and change-makers at a conference which I attended in Stockholm. She spoke just before former President Obama, and received rapturous applause. We all know flying is bad for the environment, and yet society continues to do more and more of it – the number of airline passengers has doubled since Greta was born in 2003, and by 2037 is estimated to double again.

Flying on jet planes, it is tempting to think that the plane was going to take off anyway, and remain conveniently unclear of the actual impact. In 2016 two climatologists published a paper trying to make the impact of flights more real: one passenger's economy seat in a 2,500-mile flight (such as LA to NY) will result in the melting of thirty-two square feet of Arctic summer sea-ice cover.

163

Thirty-two square feet of ice. My mind raced to the footage of a walrus falling, its thick slabs of fat and skin bouncing and rippling like jelly, off the sharp rock edges as it fell to its death. David Attenborough explained that as there is less ice cover for the walruses (and polar bears) to stand on, they are climbing higher and higher into the rocks. My daughter watched mesmerized, then asked to rewind and watch it again, as I wondered if I should censor the brutal realities of our living moment from her.

Put simply, Greta is right – we should stop flying. After hearing her speak in her home town of Stockholm, I attempted to make the trip back to London by train. After a few hours' complicated research, and swallowing the five times higher price, I found a route via Copenhagen, Rødby, Hamburg, Cologne and Brussels. According to EcoPassenger.org that journey by train emits a third of the CO_2 emissions of the plane (or a sixth when you factor in other greenhouse gases).

It began as a deeply enjoyable experience, offering the chance to see more places, and the time to write and look out of the window. For many people, trains ask for a luxury of time that is not affordable, but when you can make the time, it is a privilege to feel life move more slowly. Then, after two days, four trains, a bus ride, a ferry crossing and a night sleeping in Copenhagen, I missed my last connection.

Three sweet Germans tried to help me find a way to make it to Brussels in time for the last Eurostar, but it was impossible. I started googling hotel rooms. 'Well, you could catch the last flight from Cologne,' one man ventured. 'Hotels in Brussels are pretty expensive.'

It was well-meaning advice and, amongst a roller-coaster of conflicting emotions, I took it. I went to the airport, offset my (very cheap) flight via Climeworks (more on them later), and cried. I cried because I felt like a failure, and this seemed to represent a wider, deeper failure in our systems that makes flying so easy, and so much cheaper, than the alternatives. At the airport, they insisted I buy an ever-lasting clear plastic bag to put all my toiletries in. I read about a hungry polar bear entering a Russian town for the first time in forty years, and Heathrow's planned new runway, before getting into bed by midnight.

Whilst advocating train travel – which I still do – it was perhaps an important experience to have, because there is a reality to be faced around our existing travel choices that perpetuates the demand for flying. Often trains are great: communal spaces where you can people-watch, have surprising conversations, or lose yourself in the view. I have relished travelling through Canada, the US, India, Japan (ah the Bullet!) and Europe by trains, reading as the world sweeps by. (*Tagskryt!*)

Yet trains are not always easy to coordinate, especially across long distances. And how can we expect an absolute movement away from flying, and towards trains, when they are often harder to book and more expensive?

Unfortunately, as soon as you make the decision to leave your town, travelling becomes obscured by

Train bragging. Whilst I'm at it: the bunk beds, magically folding out from the grey-felt-lined walls of a 1980s Hitchcock-ian sleeper train, provoked an un-paralleled squeal of delight from my daughter.

shades of grey. For example, it turns out that driving solo by car (unless it is hybrid or electric) emits more than flying, especially as the fuel efficiency of planes out-paces that of cars. Meanwhile, planes also release vapours high in the atmosphere, the implications of which are not fully understood by scientists.

I respect and commend the commitment of people who choose not to fly, and I hope Greta will inspire many others (as she has me) to further reduce our carbon footprints, and radically question our choices. Yet, whilst the amount of air travel continues to sky-rocket, the pragmatist in me wants to think seriously about deep counter-trends we might try to invoke: better, faster, cheaper and more distributed trains (calls for renationalization abound); electric buses, electric planes and boats, affordable sailing (if you're not a sailor), offsetting or frequent-flyer taxes.

Flying only contributes 2.4 per cent of global emissions (compared to road transport's 16 per cent), but it represents a distinctly elitist form of pollution: 15 per cent of the UK population account for 70 per cent of the flights. If you fly, it's likely to be the most carbon-heavy thing you do.

If we are not to ban flying (which usually proves a bad policy: think Prohibition or the War on Drugs), can we align freedom with fairness? The campaign A Free Ride advocates that everyone should be allowed one tax-free return flight a year, and then be taxed from the second flight onwards, at a rate which would increase the more they fly.

What did Americans do in response to the 1920 constitutional ban on drinking? They carried on drinking, through loopholes in legislation or by turning to the black market of bootleggers and speakeasies. Although studies have indicated that the overall level of drinking did reduce during Prohibition, along with a reduction in liver disease, it was also tied to a boost in organized crime and a negative impact on the economy. It was eventually repealed in 1933.

Leo Murray, the founder of A Free Ride, told me, 'In short, *nobody has any better ideas* for what to do about the fact that demand for flights is growing much faster than technology is improving. We have to constrain demand for air travel, but if we do that with a blanket price increase then we'll price the poor from the skies altogether. This way is radically fairer and also more targeted and effective.'

The proposal has been supported by a wide range of environmental organizations and unions and was recommended to the UK government by the Committee on Climate Change. The revenue could be used to fund environmental restoration; or (Leo's preference) to support greener alternatives to flying. Can flying green itself?

When the ozone hole was discovered, we didn't tell everyone to give up having a fridge – instead, we pushed the industry to build better fridges. What would better alternatives look like in aviation? Electric planes, bio-fuels, or perhaps new visions entirely?

In 2011, forty years after President Nixon launched the US government's aggressive, costly and globe-spanning war on drugs, a report by the Global Commission on Drug Policy opened with the damning line, 'The global war on drugs has failed, with devastating consequences for individuals and societies around the world.'

Electric dreams

Why, sometimes I've believed as many as six
impossible things before breakfast.
Lewis Carroll, *Alice Through the
Looking-Glass*

In 2015 the Swiss Solar Impulse 2 plane made the first fully solar-powered circumnavigation of the globe.

In 2019, a $4 million nine-seater electric plane, called *Alice* – able to fly 650 miles at 500 mph – became commercially available. Electric dreams are slowly coming true.

There are over two hundred electric aircraft in development. Wright Electrics are working with EasyJet, for example, to have a fully electric plane running a fifth of their short-haul flights by 2027. Sweden and Norway aim to make all short-haul flights electric by 2040.

Electric is already powering drones for package deliveries, especially useful in rural, poor communities where road access can be limited. In Malawi the UN used drones to deliver HIV tests for early infant diagnosis.

Electric flying taxis are also becoming weirdly real: in 2019 a German start-up tested a five-seater electric flying car; and other industry giants, such as Airbus and Uber, aim to have commercial electric flying taxis operational in the 2020s.

Long-haul electric planes are still problematically out of reach. Hartmut Neven and his Quantum AI team are working with NASA to develop lithium-oxide batteries, which may be powerful enough to see long-haul commercial electric planes take flight.

Meanwhile, bio-fuels are being tested as alternatives to fossil fuels for aviation, with major airlines investing heavily into them. There is even a company turning household waste into sustainable aviation fuel! The California-based company, Fulcrum, say their technology can cut airlines' carbon emissions by 80 per cent compared to traditional jet fuel.

A Spacefaring Species

As counter-intuitive as it may sound, we may one day find flying to be the most energy-efficient way to travel ourselves and our stuff. Taking to the sky offers the potential to leapfrog the heavy land use and infrastructure demanded by railway lines and roads: akin to how solar panels have leapfrogged grid systems. Is this a utopian or dystopian vision of our future? Hopefully, the wilderness of our skies won't get too busy.

Recall how the symbolism of the train has shape-shifted from catalyst of coal, to eco-brag, to class-delineating *Snowpiercer*. Might our contemporary eco-nemesis, the plane, surprise us?

New visions entirely

It would be great to have an alternative to flying
or driving, but obviously only if it is actually
better than flying or driving.
Elon Musk, Hyperloop Alpha report

Travelling between San Francisco and Los Angeles in the process of writing this book, I opted for the train. We inched out of the station and wove slowly between the highway and the sea, up the magnificent Californian coastline. A red car raced next to us. For a while we rode together, then the car sped ahead and eventually disappeared into the horizon. Twelve hours later we reached our destination; probably a few hours behind the car, and long after a plane would have landed.

In 2009, a proposal was made to build a fully electric high-speed rail link between San Francisco and Los Angeles that would reduce the journey to two hours and forty minutes. $60 billion later, that line is yet to materialize: now pledged for 2029. In 2013, in response to the proposal, Elon Musk wrote a 56-page White Paper, drawing attention to an alternative that has haunted the imaginations of writers and scientists for centuries: 'Short of figuring out real teleportation, which would of course be awesome (someone please do this), the only option for super-fast travel is to build a tube over or under the ground that contains a special environment.'

The 'special environment' Elon was describing referred to an idea first posited in 1904, by the American engineer Robert Goddard. Goddard's proposal for a 'vactrain' recommended creating a vacuum in a tunnel to reduce air resistance, so that trains could move through very quickly, using less power. The idea is to essentially mimic the atmospheric conditions in space, or at high altitude, and enable travel that is competitive with the speed of planes. In the hundred or so years since then, many scientists and science-fiction writers have revisited the basic premise of this idea, embellishing it with possibilities like magnetic levitation, yet it has never been operationally realized for mass transit.

Elon's White Paper proposal for a 'Hyperloop' re-popularized Goddard's idea, arguing that the Hyperloop would be significantly cheaper, faster and safer than the high-speed electric train, making the LA–SF journey in about thirty minutes. He advocated placing solar panels on the top of the tube to make the Hyperloop entirely self-powering, and then he set up The Boring Company to start digging tunnels in California.

> To solve the problem of soul-destroying traffic,
> roads must go 3D, which means either flying cars
> or tunnels. Unlike flying cars, tunnels are weather-
> proof, out of sight and won't fall on your head.
>
> The Boring Company,
> www.boringcompany.com/faq

The Boring Company claims that it can build layered networks of tunnels, at a tenth of the normal cost of tunnelling, which will facilitate mass electric travel. It has proposed two types of travel that can be provided through these tunnels: 'loops' which would enable cars, pedestrians or cyclists to travel on 'electric skates' capable of travelling at 125 to 150 mph; and the Hyperloop which would use pressurized cabins and vacuums to transport 'electric pods' at over 600 mph.

Initial studies from NASA into the Hyperloop proposal suggest that it could provide a 'faster and cheaper alternative to short-haul flights'. A feasibility study by the US Department of Transport estimated that it could be six times more energy efficient than air travel on short routes, and three times faster than high-speed rail systems.

Hyperloop Transportation Technologies are planning to build a Hyperloop to connect Chicago and Cleveland – making the 350-mile journey in under thirty minutes – and Hyperloop One have started testing in Nevada. Other US cities and different

Floating slightly as the name implies. Although the concept has been around since the early 1900s, only a small number of commercial models have ever been rolled out. The first was a slow and steady shuttle that ran between Birmingham International airport and Birmingham International railway station from 1984 to 1995.

countries around the world, such as China and Germany, are examining the use of the Hyperloop to reduce traffic congestion, decrease noise and air pollution, and move into a more environmentally friendly age of mass transportation.

Meanwhile in China, plans for a <u>magnetic levitation train</u> have been proposed, to go into production in 2021, which would travel at 373 mph – making the trip between Beijing and Shanghai in three and a half hours – less than the four and a half hours it currently takes by plane. In Japan, magnetic-levitation trains already reach speeds of 375 mph. A group of Chinese scientists has developed a proposal to build a vacuum tunnel under water, which they say would enable carriages to reach the mind-blowing speed of 1,240 mph.

A spacefaring species

On this day I climbed a tall cherry tree at the back of the barn . . . and as I looked toward the fields at the east, I imagined how wonderful it would be to make some device which had even the possibility of ascending to Mars.
Robert Goddard in *Robert H. Goddard: Pioneer of Space Research*, by Milton Lehman

Whilst waiting for electric planes and Hyperloops to arrive, travel – and flying in particular – asks important philosophical questions of us. How do you square human innovation and the cultural benefits industrialism has given us, with the environmental toll it takes? How do you tell a community that depends on tourism (perhaps even eco-tourism) for their livelihood, that most of it is not OK?

From a fundamental philosophical perspective, is it right to put the lid on human innovation? Should we have discouraged Leonardo da Vinci from dreaming of helicopters, Robert Goddard from dreaming of space travel, Amelia Earhart from flying solo across the Atlantic Ocean, Neil Armstrong from walking on the Moon, and the companies now trying to reach Mars?

Was it *bad* that I flew to Brazil? Is it *good* that the indigenous tribe I met when there have started flying around the world to spread their knowledge? Is a helicopter fighting a forest fire or running a children's ambulance service a *good* helicopter? Is a helicopter taking tourists sightseeing a *bad* helicopter? Can we get one type without the other?

Indeed, paradoxically, might flying be the very thing that could save human civilization long-term by enabling us to become a multi-planetary species. In 2017, Stephen Hawking publicly voiced support for colonizing the Moon, Mars and Proxima B – the habitable planet that scientists believe exists in our nearest star system. He argued that 'to stay' on Earth alone 'we risk being annihilated . . . Spreading out into space will completely change the future of the human race, and maybe determine whether we have any future at all.'

I had gone to interview Elon about his adventures on planet Earth but we quickly spiralled into

Why use a rocket once when you could use it twice? A bit like a reusable water bottle, but for rockets.

A scientific research facility in Arizona, designed to be sealed off from the outside world and emulate the environmental conditions of planet Earth. In the 1990s volunteers attempted to live inside the biosphere for years at a time to assess whether humans were capable of living in self-sustaining colonies in outer space. The missions, while groundbreaking, were beset by low levels of oxygen, continual hunger, infighting and plagues of ants and cockroaches.

a black hole discussing the practicalities of space travel. Elon says that when he founded SpaceX – a company dedicated to space exploration, 'reusable rockets', and the colonization of Mars – the 'defensive' argument was an important consideration: 'that it's important for humanity or life as we know it to be on many planets, in case something happens to one of the planets, [so] the light of consciousness will not be extinguished'. Yet his main motivation spoke more to the human spirit of adventure: 'I think it would just be a far more exciting future if we're out there among the stars than if we are for ever confined to Earth.'

Might honeymoons to the Moon be a norm for our grandchildren?

Elon is not alone in his mission. Multiple companies are competing to make space flights commercially viable – World View (a floating helium balloon, founded by the scientist couple who ran Biosphere 2) and Virgin Galactica among them. Amazon CEO Jeff Bezos is funding a space company, Blue Origin, with the vision of moving 'big factories and pollution-generating industries' to space, so that 'Earth can be zoned residential.'

Whilst extremely excited by the adventurous possibilities (I've always wanted to space-travel), I also feel deeply suspicious of conversations about space colonization: its name offers an unabashed extension of the same mindset that saw (predominantly white male) Europeans ravage and plunder new continents. So far, only white men have visited the Moon; and it

happens to be wealthy, white men leading space ambitions now.

It also seems a distraction from the more important task at hand: bringing balance to our
unique planet. In 1969, Ralph Abernathy, succes-
sor to Dr Martin Luther King Jr as president of the
Southern Christian Leadership Council, led a protest
alongside the Apollo 11 Moon mission, saying, 'We
may go on from this day to Mars and to Jupiter and
even to the heavens beyond, but as long as racism,
poverty and hunger and war prevail on the Earth, we
as a civilized nation have failed.'

Elon reckons that SpaceX will be able to make the
first Mars landing in the next four to five years; but to
develop a habitable colony, 'we're talking efforts that
are across the span of centuries, a century at minimum'.

The trip would take three months one way – 'better
than Australia in the old days' – and the colony would
aim to be self-sustainable, through mining and planting, so 'if the ships from Earth ever stopped coming
for some reason, people could still survive'.

How would SpaceX ensure diversity of people
on Mars? I ask. Elon referred to indentured servi-
tude, government sponsors or simply buying tickets
– joking that the return ticket might be much more
expensive than the outbound.

How would they select which animals would go
out on Elon's Ark?

'We'll send all the animals that we want – cuddly
rather than spiky ones,' he laughed, 'and probably
some will come along that we don't want.'

One often touted form of geo-engineering is solar radiation management, whereby sunlight is reflected and global warming is reduced.

175

When nineteenth-century astronomers observed Mars using early low-resolution telescopes, many claimed to have witnessed a network of long straight lines. After some debate, the possibility that they were irrigation canals, built by intelligent life forms to transport water from the planet's melting polar ice caps to more arid regions, was agreed to be the only logical explanation.

Would he go? 'If I'm alive I'll do it.'

Elon and his twelve-year-old son Kai then launched into the serious practicalities of changing the atmosphere of Mars, ringing familiar of geo-engineering conversations, or the nineteenth-century belief in Martian canals. We could use plants to create more oxygen on Mars, and warm it up by 'terraforming' it, so that, eventually, humans might be able to walk and breathe out in the open air.

'There are slow ways and fast ways to warm it up,' Elon said. 'The fast way would be a series of giant thermonuclear explosions centred over the poles.'

Kai asked, how would you do it slowly?

'Polish the Moons and turn them into a big reflector.'

'They'd have to be colossal – that just doesn't seem feasible,' his son replied.

'You see, Kai knows his science!' laughed Elon. 'There are a few ways to skin the cat, and that's one of them – but that would take a long time. Whereas thermonuclear power, that's easy.'

Down to Earth

In the sciences we have learned that we are too ignorant to safely pronounce anything impossible.
Robert Goddard, *The Papers of Robert H. Goddard, Vol. 1: 1898–1924*

Greta and Elon are emblematic of the diverse spectrum of philosophical positions when it comes to thinking about the environment: do we need to think down (to Earth) or look up (to the stars)? A Prophet insists on behaviour change that is grounded and humble; a Wizard insists on innovation that is pragmatic and ambitious. We likely require both.

In discussing travel, I have not explored a different type of passage that is becoming increasingly necessary: climate migration. It is a painful paradox – and climate injustice – that the act of some people increasingly travelling (high-carbon) for leisure, is driving more and more other people to move by necessity.

In the first half of 2019, 7 million people were internally displaced because of extreme weather – twice the number of those displaced by conflict and violence – yet 'environmental refugees' do not have official refugee status. The World Bank predicts that by 2050 there will be as many as 143 million climate migrants.

So, alongside encouraging companies and politicians to enable cleaner modes of transport, we also need to encourage sympathetic political structures and border policies, that can help deal compassionately with the type of travel no one asked for.

So you find a nice new planet and it looks lovely. Maybe you want to move there. Sure, it looks pleasant enough from far away, but what if it's too hot? Or too cold? Or simply covered in poisonous gas? The wonderful, and still entirely theoretical, answer to this problem is terraforming, where a celestial body can be altered in such a way as to make it habitable for humans and other Earth-dwelling life forms.

Chapter Seven
Playing God or Re-Wilding

Geo-engineering and the miracles of nature

I had the opportunity to build a house for
myself. When I was done, it was time to put in
the hedges and I waited for the rain to plant.
I waited, and I waited. I waited and I panicked.
I waited and I was sad.

Mary Fernando Conrad

'The rain didn't come that year. Later in the year, things began to die. The landscape changed. Big trees became big skeletons. It was droopy, grey and barren, lost and creepy to look at, a burnt-out house, a cinder. I gave up waiting and I mourned. One day I was standing at the back of the house and I looked up at the solar panels. I was still mad, perplexed, and I just thought: why can't those solar panels also take in carbon dioxide or whatever is causing this problem?'

In the basement of her home-turned-studio in uptown San Francisco, Mary Fernando Conrad is pinning a thick yellow wire to an old drawer. She is getting ready for her upcoming exhibition, titled 'What is Shakespeare in a Time of Climate Change?' in which she plans to display a 'gismo that removes carbon dioxide' from the environment.

Will the gismo actually work?

Mary has been trying for a few years to engineer something *functional* – a domestic appliance, tile, surface or paint – that will absorb atmospheric carbon dioxide, but it turns out that is a tall and expensive order.

Instead her sculpture speaks to a nascent boom of interest in machines capable of removing carbon dioxide from the environment. Can technological wizardry save the day? Will robot-run laboratories, Artificial Intelligence or carbon-dioxide-sucking machines fix the problem? Can technology play God?

In their 2018 report, the IPCC called for 100 billion to 1,000 billion tons of carbon dioxide to be removed from the atmosphere this century. Researchers and entrepreneurs are vying for solutions. This fits into a broader investigation, often called 'geo-engineering' – large-scale, technical (but risky) ways to 'solve' climate change.

United Nations Intergovernmental Panel on Climate Change. Basically, the people we really need to listen to.

Whilst we were discussing Mary's ideas, my daughter ran outside and started watering the ferns which trembled gently in the breeze. 'It's a beautiful garden!' she exclaimed. It was. Moving my attention between her and the gismo, offered an apt metaphor for the paradox techno-utopia finds itself in: whilst it is hugely invigorating (and increasingly necessary) to consider carbon-capturing machines and human-made solutions, are we overlooking the much simpler solutions on our doorsteps: ferns, plants, trees, soil – all the miraculous technology of nature itself?

Carbon-sucking machines

> Yea, all which it inherit, shall dissolve,
> And, like this insubstantial pageant faded,
> Leave not a rack behind. We are such stuff
> As dreams are made on.
>> Shakespeare, *The Tempest*

We are made of carbon; we breathe it, burn it and dream of other carbon beings. As carbon dioxide boils the bath we live in, carbon-sucking machines have lit up the imaginations of environmentalists, capitalists and the fossil fuel hungry alike, and emerged as a foetal reality.

Klaus Lackner at the University of Arizona was the first person to explore the removal of carbon dioxide (CO_2) from thin air – called 'direct air capture' – in the 1990s. He built an 'Artificial Tree' out of a resin that absorbs CO_2 at one thousand times the rate of nature, and advocates building 100 million of them to match the amount of CO_2 the world currently emits annually.

In Switzerland, Climeworks operate the world's first commercial direct air-capture plants. Their machines use turbine fans, powered by waste heat, to filter thousands of tons of CO_2 annually from ordinary air, turning it into a concentrated form sold for commercial use, or pumped underground.

Dan Nocera, a Harvard chemist, has developed an 'Artificial Leaf' which replicates photosynthesis with ten times the energy efficiency of nature; and a 'Bionic Leaf' which can produce carbon-neutral fuels and fertilizer. He claims his technology can create cheap, off-grid, solar-powered and carbon-neutral fuels from sunlight, water and air, whilst extracting carbon dioxide from the air in the process.

Nocera has had a hard time gaining traction for his products in the West as the 'trillions of dollars' that we have already invested in fossil-fuel power technology make it hard for new fuel products to compete economically. Instead he has decided to focus his efforts on the lower-income world, working in partnership with a university in India, where there are fewer 'sunk costs' to overcome. 'People say, it's nice you are helping poor people,' Nocera explains, 'and I say, no, poor people are trying to help me.'

How much is saving civilization worth?

We should keep in mind that these initiatives
to suck carbon out of the air are in early
development and look at them with curious
scepticism. There's a lot of air out there and it
takes a lot of energy to suck it and clean it.

Lisa Jackson

The technology for removing carbon dioxide from the atmosphere exists, but there is much scepticism around it because there is a lot of air in the world to filter, so it doesn't come cheap. Capturing CO_2 at a factory costs approximately $80 per ton, but capturing it from thin air – the scale we would need if we were trying to 'fix' global warming – costs as much as $600 to $800 per ton.

'Even if the quantum computer gives you the perfect solution for replicating photosynthesis it may be the case that it is too expensive to implement at large scale,' says Hartmut Neven. Matt Lucas, from the Center for Carbon Removal, is more optimistic that 'technologies to capture CO_2 from the air have the potential for the sort of steep price declines that we've seen from solar, wind and batteries'. Meanwhile, Canadian company Carbon Engineering claim they have managed to reduce the cost of direct air carbon capture to $100 a ton.

That would still require a $10 trillion to $100 trillion bill to capture the levels of CO_2 recommended by the IPCC, which governments are only likely to go so far to meet.

Thus, making a commercial case for CO_2 capture is essential to the success of this burgeoning industry, both to financially support development, and to find new homes for the CO_2 to physically live in. Can we turn it into something useful? Back to alchemy.

The desire for fuel is largely driving demand. Silicon Valley company Opus 12 have created technology that can turn any source of CO_2 into a gas akin to natural gas. The IPCC's favourite form of CO_2 removal technology would involve growing crops (which naturally absorb it), burning them for energy, then capturing the CO_2 released and storing it, likely underground. Called BECCS, this process is already happening on a small scale around the world.

However, bio-fuels require a lot of arable land – competing with food – and one review estimates that to make BECCS work at the scale needed, it would require land three times the size of India.

Meanwhile, CO_2-capture technology is getting a surprising boost from the oil industry. A collision of disparate interests saw an unusual law come to pass in 2018 in the US. The legislation, which offered tax credits to make it cheaper for companies to retro-fit fossil fuel power plants with carbon-capture technology, had originally been promoted by climate-change deniers, because captured CO_2 can be pumped into the ground of unproductive oil wells, to help push up and extract more oil. The legislation was backed by environmental NGOs keen to see this technology advance, and finally signed into law by President Trump's government.

Construction is also a good place to focus innovation, given that concrete is, after water, the most widely used substance on Earth: if it were a country, it would be the third-largest emitter of carbon dioxide after China and the US. Construction also offers large areas within which to sink CO_2 – which we need if we plan to capture lots of it.

So, can we make architecture a solution rather than a problem? That was, you might recall, Apple Park architect Stefan Behling's vision. Assuming that the trees are replaced, wood offers a natural carbon sink to build with. Stefan Behling tells me that mass timber is now increasingly being used, in place of steel or concrete, for multi-storey buildings. He also advocates green roofs for natural carbon sequestration: if you put a 'green wig' on top of a building, 'the bee or the bird or the worm doesn't know it's a wig. In a perfect world, every square inch of this planet would be covered with organic natural matter.'

Other architects are asking, can we turn CO_2 into bricks? At the University of Aberdeen, engineers have developed technology to convert captured carbon into products such as plastics, papers, glues, paints, cements and concrete.

Meanwhile, also in the UK, Ecotricity founder Dale Vince has started mining diamonds from the air. The Sky Mining Company's factory in Stroud turns environmental CO_2 into diamonds. As the factory is run by renewable energy, Dale maintains, 'The air we put back in the atmosphere is cleaner than the air we take in.' He thought of diamonds because they offer 'probably the most permanent form of carbon capture – you can't break them'. There is not a lot of carbon in a diamond, and so this will only have a big impact if they can make 'giga amounts'. I immediately imagine a Diamond Age, with glittering furniture, buildings and even cities.

Indulgence

We need a revolution in consciousness.
                                        Dale Vince

Wouldn't it be cool if there was someone taking CO_2 out of the atmosphere and locking it down? If we did that every time we flew, maybe we wouldn't have to feel bad about flying?

Well, that's happening, and you can buy into it.

As I went to offset my flight in Cologne airport, after my failed train trip home from Stockholm, I discovered the first consumer carbon-capture product. Climeworks have partnered with CarbFix to capture CO_2 then pump it underground in Iceland, where it mineralizes into the porous basalt stone after less than two years. Basalt stones are common around the world, on oceanic floors and 10 per cent of continents so, 'in theory, basalts could permanently hold the entire bulk of CO_2 emissions derived from burning all fossil fuel on Earth,' says Sandra Snæbjörnsdóttir from CarbFix.

Climeworks hope the project will help them meet their stated goal of capturing 1 per cent of global emissions by 2025. They are offering the first consumer carbon-capture product by way of a subscription service (€8 to €49 a month) to pay to turn CO_2 into stones: marketed as a more tangible way for individuals to offset their travel emissions.

The stones are currently much more expensive than conventional offsetting: my flight from Cologne to London cost £80. Fully capturing the CO_2 from one passenger's contribution to a round-trip flight between New York and London would cost $1,400 in economy ($5,000 in business). These prices will drop as the technology develops (recall that solar panels

At €49 a month, you can feel content that you have removed 100 per cent of the global average travel footprint per year. (Although if your conscience has taken you to subscribe, your footprint is likely bigger than average.)

have dropped 99 per cent in recent decades), but meanwhile, might make us also pause to consider the true cost of our trips.

Carbon offsetting has always reminded me of that episode of the TV show *Friends*, when Phoebe (the hippy vegetarian) became pregnant and asked her friend Joey (the not-so-smart, sweet actor) to be a vegetarian in her place, so she could eat meat. Offsetting works by trying to prevent someone else from emitting carbon dioxide (through stopping deforestation, planting new trees, methane capture or supporting renewable energy projects), equivalent to at least the amount of CO_2 your action has created.

Only 1 per cent of passengers currently offset their flights, but the number is growing rapidly: with 140-fold growth between 2008 and 2018. Meanwhile, CORSIA is a UN agreement, supported by 192 countries, that mandates that any aviation emissions above and beyond 2020 levels will be offset by airline companies.

In the fifteenth-and sixteenth-century Netherlands, the absolving of one's sins had a precise tariff. One could receive a godly pardon for poisoning for eleven ducats and six livres tournois, while incest would set you back thirty-six livres and three ducats.

Offsetting is an imperfect, philosophically dubious solution, that risks dangerously appeasing our consciences on carbon-gluttonous choices like flying. It is tempting to compare it to 'indulgences' in the Middle Ages, when people could pay priests to absolve them of their sins. It risks creating a moral hazard, encouraging people to participate in a business-as-usual approach to emissions, when we actually need to *both* reduce our emissions, *and* invest in nature.

'We need to maximize both the decarbonization of our economies and natural climate solutions. We cannot play one off against another,' George Monbiot writes to me on this point. 'It is arguable that offsets were never appropriate. But if they were, that era has now passed.' We now live in a world where *both* Joey and Phoebe (and Rachel, Ross, Chandler, Monica and Gunther) need to stop eating meat.

That said, the pragmatist in me kicks in. People, companies and politicians still fly, and emit greenhouse gases in a million other less obvious ways. It is very hard to have a zero-carbon footprint. So, after reducing our carbon footprints as much as possible, and short of electric planes or a frequent-flyer tax (which would act as a mandatory offset), *voluntary offsetting* is better than *not offsetting*. It also helps to ensure that natural climate solutions are better funded, when it is estimated that only 2 per cent of philanthropic spending goes towards climate change.

I have offset personal and company flights through the World Land Trust (WLT) who focus on protecting existing forests and wildlife corridors: 'We try and let people know that everybody has an impact, regardless of whether we like it or not. We encourage people to measure, reduce and offset,' says Dan at the WLT. Meanwhile, Dale Vince – who is also not convinced by existing offset schemes – is hoping Sky Diamonds will appeal to people wanting to be certain CO_2 has been removed from the atmosphere. They can do so by taking home a 'sack of diamonds'.

Just like corridors you might find in a building, helping people get neatly from one place to another, but made of trees and bushes. A clever way to join together disparate conservation areas, following animal migration paths, so you achieve more wildlife protection using less land, and minimize animal/human conflicts.

Hack the planet

Do polar bears have rights?
David Keith, *A Case for Climate Engineering*

Carbon-sucking machines are only the tip of the iceberg. As scientists tell us that the Arctic is warming twice as fast as everywhere else; that Greenland's ice-sheet is already melting four times quicker than it was ten years ago; and that it is anticipated that by 2030 the Arctic will be ice-free in summertime, others wonder if we can re-freeze the polar caps.

Geo-engineering was first proposed in 1965, by US President Lyndon B. Johnson's Science Advisory Committee, who warned of the dangers of fossil fuel emissions, and suggested researching counter-measures including 'raising the albedo, or reflectivity, of the Earth'.

Perhaps we should decorate and whiten the Arctic with light-coloured aerosol particles, or spray seawater into the atmosphere above it to create clouds that reflect more solar radiation back into space?

Or we could feasibly place 10 million wind-powered pumps over the Arctic ice cap, to thicken it by pumping water to the surface to freeze, at an estimated cost of £400 billion or $500 billion – to put things in perspective, that's less than the annual US military budget.

Maybe it would be a better idea to blast sulphates into the atmosphere, mimicking a large volcanic eruption like that of Mount Pinatubo in 1991 which cooled the planet half a degree?

Or what about putting shields and satellites in the sky, like a giant sun umbrella?

How about we scatter iron powder in the oceans, to increase the amount of phytoplankton, and create 'algae blooms'? Or mine olivine, or sand down rocks, and disperse the minerals across agricultural fields and oceans, because they are good at absorbing carbon dioxide? Maybe we could genetically engineer plants to improve photosynthesis?

If all else fails, build a wall.

Not just any wall. An underwater sea wall, to trap and catch glaciers before they slide and melt into oblivion.

> A bad idea whose time has come.
> Eli Kintisch, *Hack the Planet*

These are the ideas being seriously, sombrely and increasingly debated by scientists and <u>technogaians</u>, looking for ways to 'hack the planet' in order to avert the escalating climate crisis.

'We may already be at the stage where some of the effects of climate change are effectively irreversible unless we do something really dramatic to try to "restore" them,' climate scientist Emily Shuckburgh tells me. Emily helped found the 'Climate Repair' team at Cambridge University who research and evaluate different 'emergency measures' including geo-engineering proposals, and their potential negative side-effects such as droughts arising from changed rainfall patterns and damage to ecosystems. She is

Marrying Tech and Gaia, this breed of environmentalist embraces technology to help pave the way to a cleaner, greener future. Fraser Clark was a technogaian who advocated balancing the 'techno right brain' with the 'hippy left brain' and organized consequent 'zippie picnics' in 1980s and 1990s London.

sceptical of geo-engineering but recognizes that 'the world is clearly in somewhat of a crisis situation, and we don't want to rule anything out'.

Artificial Intelligence and quantum computing may also help model the possibilities for solar engineering, and if we reach unforeseen future tipping points – black swan events – we might need technical solutions so immediate and complex that only AI could imagine them. 'I'm not sure to what degree I like these ideas,' says Hartmut Neven, 'but given the timelines we are looking at, if we can't reduce CO_2 emissions fast enough, then we may have to do something like climate engineering.'

In the half-century since geo-engineering was first posited, it has largely split the environmental community. David Keith (founder of Carbon Engineering) and James Lovelock advocate further research; David Attenborough has equated it with fascism and David Suzuki calls it insane.

> If another civilization was looking at us they would say, wow, they're really putting a lot of effort into taking carbon out of the ground and putting it into the atmosphere and the oceans.
>
> Elon Musk

Geo-engineering scares me. But I'm already scared. We began our geo-engineering experiment a long time ago, when we started digging up and burning the Earth's underbelly. We are already manipulating our planet on an unprecedented and unpredictable scale.

For a very long time in the European imagination black swans (the animals) were thought not to exist and the phrase was used to describe something impossible. That was until they were 'discovered' in the late seventeenth century by European explorers in south-western Australia. The term has since come to define rare surprise events that are paradigm shifting, with widespread ramifications (both good and bad) and often lead to attempts to rationalize them in retrospect, such as the rise of the Internet, or the First World War.

Some countries are already using planes to spray salts into clouds – <u>cloud-seeding</u> – to enhance rain.

Research is sensible, in case we do reach sudden climate tipping points (and they may prove strangely handy in the next naturally occurring <u>Eocene</u>), but further geo-engineering offers us a last resort. These ideas won't fix multifaceted aspects of climate change like ocean acidification, and they may cause unforeseen consequences. They may help certain regions at the expense of others, and introduce new risks. They offer to Band-Aid our problems, rather than drive to the root causes, and encourage a business-as-usual mindset to emissions and mass consumption.

I realize, resisting the Band-Aid, what a perversely wonderful opportunity our climate crisis actually offers us. Sometimes the house has to crumble so you can build a new and better one. Perhaps this crisis is in fact a blessing – an opportunity to make transformative change.

If we don't want to geo-engineer our planet any more, can we step back and let nature fix it?

Withdraw

I went to the woods because I wished to live deliberately, to front only the essential facts of life, and see if I could not learn what it had to teach, and not, when I came to die, discover that I had not lived.

Henry David Thoreau, 1854, *Walden*

The country with the most extensive cloud-seeding programme is thought to be China, particularly over arid regions, where silver iodide rockets are regularly fired into the sky to increase precipitation or snowfall. Famously used to clear the skies for the Beijing Olympics in 2008.

Fifty-six million years ago, crocodiles' and turtles' ancestors swam amongst polar forests, on a planet that was pretty sweltering. Over the next 22 million years, Antarctica went from subtropical rainforest to the ice sheet we anxiously pine to preserve today. Scientists are not entirely sure what exactly caused the extra carbon emissions that triggered *(cont overleaf)* 'hot-house' Earth (volcanic activity? A comet impact?) but comparisons are drawn with our dawning predicament today.

A leaf from Walden Pond, the location of Henry David Thoreau's experiment in solitary living in nature, was stuck to the wall of Mark Boyle's handmade cabin in Ireland, before my daughter reached to touch it and it fell down. Mark picked the leaf up tenderly, and told me not to worry – he would put it back later.

After gaining some fame for writing a book about living without money for three years, Mark set up a free hostel in Ireland, hand-built his own cabin and in 2015 began living without electricity, modern technology or running water. No fridge, freezer, lights, washing machine, phone, TV or computer. Not even a pen.

Mark's cabin is filled with books like *Confessions of a Recovering Environmentalist* by Paul Kingsnorth. His waste bin, collected over five months, contains less than mine does after a week. Great piles of wood for heating and cooking line all the walls inside and out, and candles burnt down to their core decorate the room.

Mark grows all of his food, cycles miles to a local lake to fish, and collects water from a spring nearby. Friends and strangers regularly drop by to stay, drink from his free bar, or offer a helping hand.

> Withdraw because refusing to help the machine advance, refusing to tighten the ratchet further, is a deeply moral position. Withdraw because action is not.
> Paul Kingsnorth, *Confessions of a Recovering Environmentalist and Other Essays*

Mark doesn't buy the tech-will-save-us narrative. He is concerned about the unwanted social consequences of our digital age – surveillance, addiction, mental health, fake news and cyber/AI warfare. 'Could I fall for a machine? I hope not. But I hear that some people already have, while many now spend more time in bed touching and playing with their smartphones than their lovers,' his pencil letter reads. Yet his main rationale for withdrawing from modernity as we know it, was environmental.

'In the production of one solar panel I can tell you how to destroy an entire planet, between the plastics, the minerals, and the factory system, right down to distribution around the world. Each step in the chain of making a solar panel system relies on an entire global network and a level of technology that I find inherently unsustainable and violent to the natural world.'

I haven't asked Mark what he thinks of building hundreds of millions of artificial trees and bionic leaves, because I know his answer. All the mining and metal that those artificial forests will be made from, are, in his view, the heart of the problem.

The heart of the solution

Miracles abound! We have discovered a machine already extracting carbon dioxide from the atmosphere. It gives back oxygen, fruits, food, rubber and fabrics. It offers sanctuary to birds, insects and animals like us. Just standing near it, you are likely to feel better. It gives hugs, and it is practically free.

I am, of course, describing that miraculous machine – the humble tree. Whilst we panic and scramble around, spending tens of billions of dollars, making bets on magical technical silver bullets to deal with the Sixth Mass Extinction, nature is abundant with simple and powerful solutions.

Research by the Nature Conservancy and fifteen other institutions claims that using only cost-effective solutions, nature has the capacity to absorb 11.3 billion tons of CO_2 by 2030 – that's the equivalent of stopping burning oil globally.

After discussing the wild ideas for geo-engineering with me, Emily at Cambridge's Climate Repair team concluded, 'Nature-based solutions are the way to go. The simplest possible thing you can do is plant more trees.'

Lost forests

According to the IPCC, if we do not use any carbon-capturing technology or manage to significantly curb emissions, by 2100 we will need to re-forest an area of land twice the size of Argentina or ten times the size of Spain in order to stay within 'safe' climate boundaries.

This may sound large, but when faced with the existential threat of mass extinction, and the uncertainties of techno-fixes, mass tree-planting starts to seem sensible. Re-foresting also offers multiple other benefits: protecting bio-diversity; and insulating against desertification, dust storms and air pollution.

And mass tree-planting is under way. In 2019 NASA announced that the Earth looked increasingly green from space, driven in large part by China's 'Great Green Wall' which has seen 50 billion trees planted since the 1970s; and by India, where in 2017 volunteers planted 66 million trees in one day. The twenty-year NASA study coordinated by satellites found that the world is 5 per cent greener than it was in 2000: an area equivalent to the size of the Amazon.

In 2019 Ethiopia broke the world record, planting 350 million trees in one day, as part of a wider plan to plant 4 billion. It joins a list of African nations who together have pledged to restore 100 million hectares of land. Australia has pledged to plant 1 billion trees by 2030, Pakistan 10 billion by 2023; and the Bonn challenge aims to reforest 350 million hectares of degraded land by 2030.

Swiss researchers have claimed that there is the potential to re-forest a billion hectares of Earth (roughly the size of the United States), without disrupting agricultural or urban land. Researchers claim that a forest of around a trillion trees planted on this land, at an estimated cost of $300 billion, would absorb two thirds of the additional CO_2 humanity has already added to the atmosphere.

I was rather buzzed, in Davos 2020, when the World Economic Forum launched a cross-sector initiative to plant a trillion trees by 2030. Salesforce CEO Mark Benioff thanked the usually climate-change sceptic US President Trump for backing it, saying gleefully: 'Trees are a bi-partisan issue – everyone's pro-trees.'

Three hundred billion dollars is roughly half the annual global fossil fuel subsidies, and a lot less than the costs of air capture. The price tag may get even cheaper using drones to disperse seeds. BioCarbon Engineering claim their technology is able to plant trees 150 times faster, and ten times cheaper, than traditional manual methods. They fly drones over land to map the topography and soil conditions, then shoot biodegradable pods filled with seeds and nutrients into the ground.

Tree-planting is 'available now, it is the cheapest one possible and every one of us can get involved', says Professor Tom Crowther who led the research. We can participate by supporting afforestation organizations, avoiding irresponsible companies, supporting <u>woodland burials</u> or putting our hands in the ground to plant.

Though tree-planting gets most of the limelight, there are other natural climate solutions that can, and should, also be pursued. Many environmentalists, such as George Monbiot, favour natural regeneration over tree-planting: that is, standing back, and allowing trees to seed and spread themselves (aka re-wilding), or assisted natural generation, where humans provide a subtle helping hand. Megaherbivores, such as forest elephants and rhinos in Africa and Asia, are 'natural foresters' who can help spread and germinate the seeds of trees.

Meanwhile, vegetated coastal habitats – mangroves and seagrass beds – can sequester carbon at forty times the rate of tropical forests; and peatlands

Not that anyone wants to be over-burdened with responsibility when they shuffle off this mortal coil. But it is worth considering that cremation uses about 40 litres of petrol per corpse and produces a not inconsiderable amount of emissions (which are mostly carbon but can also contain mercury from some tooth fillings). Woodland burials encourage the growth of woods and in the case of a Celtic burial, you can even have a tree buried underneath you.

contain a third of all soil carbon, even though they cover only 3 per cent of the world's land.

Protecting and regenerating these diverse natural habitats will play an essential role in efforts to stabilize the climate, yet currently, only 2.5 per cent of the funding spent on climate-change mitigation goes to natural climate solutions.

The UN estimate that the restoration of 350 million hectares of degraded land between now and 2030, could absorb one to three years of annual emissions, whilst generating trillions of dollars in ecosystem services and jobs for local communities. They have subsequently declared 2021–2030 the UN Decade on Ecosystem Restoration, as a 'global call to action' to restore our natural world.

Old forests

> Destroying rainforest for economic gain is
> like burning a Renaissance painting to cook
> a meal.
> E. O. Wilson, *The Life Scientific* (radio)

Even more urgent than planting new trees is the need to protect standing ones. It takes ten to twenty years for a tree to mature enough to be able to absorb meaningful amounts of CO_2; forests currently absorb the equivalent of Japan, Germany and the UK's combined emissions; whilst deforestation is a driving cause of species extinction, and contributes more than the world's entire transport sector.

Through the hard work of NGOs, international agreements and national commitments, 'protected areas' of land and sea are growing. Since 2018, 14.9 per cent of the Earth's land surface and 17 per cent of the oceans have been formally protected: an increase from 14.7 per cent and 10.2 per cent in 2016. In 2017 the largest protected marine area in the world was established, covering 1.5 million square kilometres of the sea alongside Antarctica.

The UN are considering a treaty to protect 30 per cent of the high seas from fishing, mining or commercial exploitation, perpetuating the oceans' role as a natural carbon sink, enabling them to better deal with stresses such as acidification and plastic-pollution, and allowing fish stocks to recover. 'It's a once-in-a-lifetime opportunity,' says Steve Trent from the Environmental Justice Foundation (EJF). 'Every second breath you have comes from the sea, in terms of the oxygen it produces.'

In 2008 the UN established a programme called REDD to provide financial incentives for lower-income countries to protect, and expand, their forests. The protection of forests internationally is also being funded through philanthropy, voluntary carbon off-setting schemes, and a $10 billion Green Climate Fund set up by 194 countries.

The World Land Trust (WLT) have protected over 760,000 acres of forest directly, and millions indirectly, by purchasing or leasing land through local NGOs to create nature reserves and wildlife corridors. Meanwhile, Rainforest Connection use technology to better protect forests: they install solar-panel powered smartphones throughout forests, which can detect and report chainsaw sounds, to help prevent illegal logging.

'There's a belief that land protection could account for 50 per cent of the carbon emissions,' says Dan Bradbury from the WLT, referring to a UN report which claims that landscapes can save more than renewables, energy efficiency and green transport put together. 'From our point of view, we believe that protecting the standing forest is the best method because that keeps the carbon that's already there locked in, and it continues absorbing more carbon as well.'

The power of no

People don't just cut down large swathes of trees for fun. All deforestation is economically driven, and efforts to stop it need to be sympathetic to the local people involved. Which inevitably loops us back around to our seemingly banal supermarket choices.

Fifty football fields' worth of tropical forests are being cut down *every minute*, 80 per cent of which is a result of agricultural production. If you knew that the beef you were eating, the leather you were wearing or the palm oil in your face cream were responsible for destroying the Sumatra or Amazon rainforest, would you still want to buy them?

'We have to tackle global commodities driving deforestation, like cattle, soya going into animal feed, or palm oil going into food, cosmetics and biofuels,' Greenpeace UK's executive director, John Sauven, tells me. Challenging consumer consumption has been at the heart of Greenpeace's work since they were founded in 1971 by a group of anti-nuclear activists.

When Greenpeace unearthed the fact that a large but little-known company called Cargill was responsible for millions of hectares of rainforest destruction to grow soya beans, 90 per cent of which were being used to feed the meat industry, they decided to go after Cargill's 'most visible client'. They launched a campaign drawing attention to the fact that McDonald's were directly driving deforestation by feeding their chickens Cargill's soya.

Through an extraordinary collaboration between unlikely players (McDonald's, the Brazilian government, Cargill and Greenpeace), and thousands of letters sent by supporters, in July 2006 a 'moratorium on soya' was established – banning the destruction of rainforest for soya production. It has since been indefinitely extended.

Through similar negative spotlighting Greenpeace also pressured Holland & Barrett to stop selling products with krill oil in them (which destroys Antarctic ecosystems), and helped to prevent Shell's Arctic exploration for oil. They continue to spotlight the commitment made by major consumer brands to end deforestation by 2020 through the 'responsible sourcing' of key drivers of deforestation – cattle, palm oil, soya and paper – which they say is failing.

'The insane thing is we are using land, not to feed people, but to feed animals, to feed cars, to feed chemical processes, all kinds of things that we cannot afford to use the land for, because we need to use that land for soaking up carbon, for biodiversity, for local community livelihoods,' says John.

Since 2017 palm oil has been a continuing focus of Greenpeace's campaigns. It is a ubiquitous product that can be found in almost half of American packaged products, from margarine to lipstick, and soy milk (as 'vitamin A palmitate') to biofuels for cars. Production has increased fourfold in the last twenty years and by 2050 is expected to quadruple again. This growth has been driven by the economic rewards it offers: a palm oil plantation returns ten times the profit per hectare of a rice plantation.

Palm oil is the principal driver of deforestation in Indonesia and Malaysia. One study estimated that from 1990 to 2005 over half of palm oil expansion in these two countries destroyed virgin forests. It has forced forest-dwelling people from their homes, and pushed animals such as orang-utans, rhinos, elephants and tigers closer towards extinction. The process also releases significant amounts of greenhouse gases such as methane, and has caused water pollution, and smoke fogs that travel as far as Singapore and Kuala Lumpur. It triggers fires and respiratory illnesses, and has been found to use child labour working in unsafe conditions below the minimum wage.

Unsurprisingly in light of all this, multiple initiatives have been aimed at improving the industry standards, including the consortium Roundtable on Sustainable Palm Oil which has certified 19 per cent of global palm oil. In 2010 Norway pledged $1 billion to Indonesia to keep its forests standing, and Indonesia officially banned deforestation in 2011, although the practice has apparently continued.

Were the true environmental costs of palm oil factored into their prices (i.e. via a polluter tax, as we will discuss in greater detail in Chapter Nine) the industry would have been much quicker and more adept at developing more sustainable practices. For example, palm oil yields can easily be increased nearly twofold through shared knowledge and this would reduce the amount of land required. It is also possible to make electricity from the methane produced by palm oil ponds: 5 per cent of facilities already do this, but if all of the refineries worldwide did, it would reduce the climate impacts of the operations 34-fold.

Meanwhile, consumers can help campaign for better business practices by supporting the essential work of groups like Greenpeace, and being mindful of the products they buy (or don't buy) into. My twelve-year-old cousin, Holly, diligently reads the packages of everything she buys and won't consume anything that contains palm oil unless it is certified sustainable.

The power of yes

At first, I thought I was fighting to save rubber
trees, then I thought I was fighting to save the
Amazon rainforest. Now I realize I am fighting
for humanity.

Chico Mendes, 1987

In the small remote town of Xapuri in the state of
Acre, Brazil, you can still see the bullet marks that ran
through the walls, when on 22 December 1988 Chico
Mendes was assassinated in the back doorway of his
home by a rancher. His wife and two children were
inside the house. His two security guards ran away.

Mendes had been rubber-tapping in the Amazon
forest since he was a child. When cattle ranchers
started clearing large areas of forest for pasture,
Mendes formed a rubber-tappers' union to protect
their rights, and the forest. The tappers involved in
this movement would stand in front of heavy machinery, using their bodies as shields: viscerally
representing the conflict between economic and ecological interests.

Mendes developed a sustainable economic policy,
called 'extractive reserves' (or 'harvesting precious
places' as Monbiot might rename them), state-owned
and protected areas of land, where local communities
could manage the land sustainably, and harvest forest
products such as acai berries, Brazil nuts, coffee and
wild rubber. Rather than crushing the rubber-tapper's
movement, Mendes's death led to a surge of support

for protecting the forest: a million hectares were designated as an extractive reserve in his name, now over 3.4 million hectares of Brazilian land function as extractive reserves.

Extractive reserves were an ingenious idea that sought to square an awkward circle: how do you align economic and ecological interests, so that the local community can make a living from wild terrain?

As urgent as the climate crisis is – and the potential for nature-based solutions to solve it – we cannot expect people simply to sacrifice their land and livelihoods. Any such suggestion risks creating a fascist *Handmaid*-like future vision, or 'fuelling environmental racism' – a charge that has been levelled against China's tree-planting schemes, for example, which apparently coerced farmers to drop their tools and start planting.

Instead, for conservation to be fair and effective, the local communities, forest-dwellers and indigenous groups need to play central roles, and have alternative ways to make a living sustainably from native ecosystems.

I had visited Mendes's home town of Xapuri, Acre, when I was travelling in Brazil, researching the potential to use wild rubber for different products – from trainers to condoms. In contrast to Greenpeace's horror stories about monolithic companies sweeping through rainforests to destroy them for crops, in Acre 80 per cent of deforestation is driven by small farms and local people, making simple economic choices.

After environmental pollution causes high levels of infertility, fertile women's reproductive rights become the property of the ruling (patriarchal, totalitarian and puritan) class. Margaret Atwood attests that everything in her dystopian novel has at some point really happened 'and therefore the amount of pure invention is close to nil'.

The Amazon forest offers very little economic value comparative to the short-term economic gains that can be made from cleared land, such as logging, arable agriculture and cattle ranching. For example, in 2012 when I visited, the average yield for native wild rubber (which depends on the forest) offered 2.5 per cent of the yield for pineapples (which requires cleared land).

Yet, based on the logic of Mendes's extractive reserves, I visited several factories and cooperatives dedicated to forest products such as Brazil nuts, timber and, of course, wild rubber.

Holding a Super 8 camera, I followed a wild-rubber-tapper deep into the Amazon forest. It was a long winding route through the trees, and I was wet with sweat by the time we arrived. I watched as he drew a rudimentary metal tool across the skin of the pará (rubber) tree, at a diagonal angle to the ground. Then he attached a little cup to the edge and we watched as the sticky milky white sap bubbled up and ran down the groove into it.

Typically, a rubber-tapper will make a series of diagonal cuts in the bark of around a hundred trees a day, spread through the forest, then return six hours later to collect the liquid latex. In a year, one tree can produce around five litres of latex. It's challenging work, with long distances covered in tropical heat and the ever-present risk of running into a snake or jaguar.

The story of the rubber industry and how it was eventually usurped by the plastics industry, reads like a fable of how capitalism has historically slashed and burnt both forests and people, yet has the potential to turn full circle to empower local communities and their environments.

Indigenous people in South and Central America had been making things from the native pará tree for millennia before the first Europeans arrived. The Mayans used rubber to make sandals, statues, waterproof cloaks, balls, rubber bands and glue. Indeed, the name of the first major civilization in Mexico – the Olmec – meant 'rubber people'.

Christopher Columbus noticed children playing with rubber balls on Haiti after his first voyage across the Atlantic, and early explorers knew about the substance, but it wasn't until the mid-eighteenth century that Europeans began investigating its uses. In 1851 Charles Goodyear and Thomas Hancock displayed extraordinary rubber mouldings – pipes, brooches, combs – at the Great Exhibition in London, before filing patents for the process for stabilizing rubber.

For many years it was the Amazon's gold, and entrepreneurs and migrant workers poured into the region. Within decades Manaus, a small town on the Amazon River, boasted more diamonds per head than anywhere in the world, grand mansions, an opera house, Brazil's first telephone system, an electricity grid and sixteen miles of tram tracks. Accounts tell of rubber barons lighting cigars with $100 banknotes and giving champagne to their horses whilst profiting from the vicious slave labour of local indigenous people.

In 1876 an English explorer and a Scottish botanist boarded a weekly steamer from the Amazon to Liverpool with 70,000 pará seeds. They took them to Kew Gardens where 2,500 germinated. In the 1880s the seedlings were sent to Singapore, heralding the beginning of Malayan plantations, and the decline of Amazonian wild rubber.

For a brief but painful moment during the Second World War when the Allies' Malayan rubber supplies were cut off, there was a boom in Brazilian production again, causing many 'rubber soldiers' to be drafted into the 'green hell' of the Amazon to tap it in the wild. Otherwise, the Brazilian wild rubber market lay dormant for much of the twentieth century: with prices unable to compete with Malayan mass-production.

Yet the recent interest in 'wild rubber' suggests that we are now living in an age when most people would prefer not to have their goods made by slaves, and in fact many would go one step further and pay more to support initiatives that put people and the planet first.

Since launching in 2005, Veja – who produce their trainers using organic cotton and use 30 per cent wild rubber in the soles – have grown exponentially: driven in large part by their positive story. Co-founder Sébastien Kopp says, 'Collecting rubber from the hevea (rubber tree) offers a better salary than cutting trees to raise cattle. Since Veja's inception, we've purchased 200 tons of wild rubber, in turn enabling us to preserve 120,000 hectares of the Amazon forest.'

The environment, and economics, continue to clash. Right now, every one to two days an Environmental Defender – like Chico Mendes – is killed for their efforts to protect the natural world. Whilst sustainability agendas, nature reserves, and offsetting schemes remain essential, the only way to fundamentally and permanently solve this tension is to fight for Mendes's vision to align economic and environmental interests.

Are the gismos simply a distraction?

What the stories of palm oil and wild rubber articulate is the subtlety with which our small, seemingly innocuous, lifestyle choices affect the landscape. Sometimes it can feel overwhelmingly confusing to navigate the impact of our consumer choices: how do you know good palm oil or soya beans from bad? How do you know which products are driving deforestation?

Some organizations are making it easier to understand: Rainforest Alliance-certified products; FSC-certified paper and wood (avoid tropical hardwood); organic labels; Stella McCartney's sustainably sourced viscose; Bottletop's deforestation-free leather, or looking for 'sustainably sourced palm oil' (as my cousin Holly does).

Another solution is pretty simple. Considering that 80 per cent of Amazonian deforestation is driven by agriculture, one of the simplest and most powerful ways to push back against this trend is to avoid meat, dairy and leather, unless you're certain of the source.

A shift towards plant-based diets would also enable a different revolution to take place: the re-wilding of our planet.

As we looked at in Chapter Two, if the world switched to a plant-based diet, the amount of land that could be freed up to give back to nature would be the equivalent of North America, Europe, China and Australia put together.

A lot of the focus of Parts One and Two of this book might read as bourgeois bullshit, to put it politely. Middle-class, New Age ways to consume better. Yet the reality is that consuming better – which involves consuming less – is essential, if we want to get serious about climate change.

Anyone spending money plays a role in directing the flow of business and production. We can simplify and resist the ideology of consumerism. Yet, unless living completely off grid, we still have to pay to feed and clothe ourselves. We still make decisions on our energy suppliers, our banks, furniture and (sometimes) how we travel. A better understanding of the impacts of these choices empowers us to drive the changes we want to see happen.

The idea for the CO_2-sucking gismo, Mary tells me, came to her many years ago when she was spending time in the wild terrain at the end of her garden.

'It's hard to measure happiness, but we can measure carbon dioxide, air quality, rate of species extinction,' she muses, leading me through the wilderness. Occasionally she bends down to pull out a <u>weed</u>. It's a feeble, crazy act in an area so vast, and I laugh.

'Well, I've just got to do what I can,' she says.

I do, genuinely, hope Mary's vision is realized, and one day we can paint and tile our homes with carbon-sucking materials. I support sensible research of geo-engineering but hope we don't need to inject the skies with sulphur, or scatter iron in the seas. Yet what I hope for more than anything is a deep shift in how we relate to – and find value in – the natural world, so that we might allow nature itself to rebound. We have the solutions to this crisis: they are right under our feet.

We can each play a part in the carbon balance of the planet: by looking at how to reduce our impact; examining what we eat and buy; supporting good ways of doing business, and boycotting bad; encouraging our politicians to prioritize natural climate solutions; and supporting carbon-capture initiatives for the inevitable excess that we can't shake, which could be as

'Weed' is a controversial term. How does Mary define it? 'It's about working with the California rain cycle (water conservation) and creating a sense of place (what is specific to here that won't thrive elsewhere) – so a mono-cropping soil-destroying water-hogger.'

simple as tending our own gardens, or as wild as buying a bag of CO_2 diamonds.

Without feeling guilty – indeed, perhaps feeling inspired by our role in this moment – we can each pull out a weed or two, to do what we can.

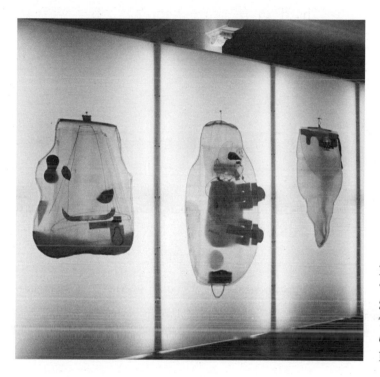

Mary Conrad, 'What is Shakespeare in a Time of Climate Change', San Francisco, 2019.

PART THREE

Citizens, Not Consumers

What does it mean to be a citizen?

> It's overwhelming to think the burden of
> keeping the world alive rests on the shoulders
> of consumers. And frankly, it shouldn't.
> Not entirely.
> Alison Stine, *Guardian*, 'Can you afford to
> be green when you're not rich?'

In the first half of this book, I explored the potential to evolve our existing systems; and how we as individuals – and consumers – might participate.

That's the doctrine by which I have largely lived my adult life: trying to work for companies I respect, founding social businesses, striving to be conscientious about what I eat, buy and how I live.

It's felt at times empowering, but also exhausting, imperfect, and inevitably hypocritical.

We cannot shop our way out of crisis.

> Grow some carrots and jump on a bike: it
> will make you happier and healthier. But . . .
> Individual choices will most count when the
> economic system can provide viable,
> environmental options for everyone – not
> just an affluent or intrepid few.
> Martin Lukacs, *Guardian*, 'Neoliberalism
> has conned us into fighting climate change
> as individuals'

Right now, the reality is that doing business in a destructive way is all too easy.

Trying to swim the other way, and produce responsible, low-impact products and services, as I have tried to do, is time-consuming, difficult and expensive.

There is also the insistent attachment to economic growth that creates inherent contradictions: you can create an ethical product, but you still have to sell ever more of them.

Meanwhile, for consumers who care, it is a bewilderingly complex landscape to navigate, and many people simply don't have the time or money to participate.

> There is a long history of industry-funded 'deflection campaigns' aimed to divert attention from big polluters and place the burden on individuals. Individual action is important and something we should all champion. But appearing to force Americans to give up meat, or travel, or other things central to the lifestyle they've chosen to live is politically dangerous: it plays right into the hands of climate-change deniers whose strategy tends to be to portray climate champions as freedom-hating totalitarians.
>
> Michael Mann, *Time*, 'Lifestyle changes aren't enough to save the planet. Here's what could'

Don't get me wrong: *conscious consumerism* remains vital in directing political and economic trends.

But it has its limits.

We need to make environmental and social respon-sibility easy, mainstream, and mandatory so that in capitalism's general race to the bottom, the bottom is not so low.

We need politics and law – the machinery represen-tative of our global consciousness – to help guide an Industrial Evolution. We need a defence and protec-tion of our commons: our air, seas and land.

The good news is, in the half of the world's coun-tries which are democracies, we have sophisticated, if imperfect, representative and legal systems, passed down by our ancestors, designed to listen to citizens.

Power lies in our collective voice, and our capacity to organize to provoke the mycelium network effect.

Ever more visionary and dedicated people are trying to generate change through working with – or sometimes against – the law, and this part of the book will be dedicated to those efforts.

Millions of children, too young to vote, and without much consuming capital, are giving voice to the future . . . and it does seem that politics and the law are, in many countries, listening.

Meanwhile, some people are trying to reform and update the political system itself.

Ultimately, we are not defined by the things we buy.

We are citizens, part of a local and global community.

What does that mean? What does it mean to take responsibility?

Chapter Eight
Power and Privilege

Protest, law and learning
from our children

As a writer, I worry that I am allowing myself to
be railroaded into offering prosaic, factual
precision when maybe what we need
is a feral howl.
Arundhati Roy, *Listening to Grasshoppers*

My anarchist environmentalist cousin Emma was
born, paradoxically, to an heir to a seat in the House
of Lords. Her father, my Uncle Adam (married to my
Aunty Jo), was a kind and gentle man, with extraordinary integrity and a face which creased up like a
ball of soft paper when he smiled. After a career as a
cinematographer he became a pioneer in sustainable
energy production. He was both an inspiration for,
and inspired by, Emma's activism.

When he was nine years old, Adam rejected the
title of Viscount, much to his father – the Earl's –
dismay. When his father later died, Adam refused his
inheritance and rejected the title of 'the Right Hon-
ourable Earl', saying he did 'not believe that anyone
should be set apart from, or above, others except by
their own efforts'.

Emma's privileged background caused conflict as
she started chaining herself to trees to protect them
against road developments: 'Sitting in a meeting, as

a teenager, listening to class war activists call me the enemy, I was confused about how I should respond,' she says.

Emma thoughtfully compares her social and political inheritance to that of her ancestry: 'One of the most difficult things about living in this world is that it is nearly impossible to avoid being complicit in its present and future destruction. We all have an inheritance that we do not want. But we have to make good, as best we can, with what we are given, even when that's not comfortable or easy.'

Inheritance

Just as we do not choose the circumstances and ancestral patterns we are born into, none of us choose the cultural norms and laws we inherit. Had my mixed-race daughter been born in a different century, she would have been a crime. If I were living in another country today, my queerness would be a crime.

So, what does it mean to be an environmentally aware person, who believes in social justice, living in a post-colonial, industrial twenty-first-century West, facing – and participating in – the Sixth Mass Extinction? Should we feel guilty? What responsibility do we bear? What forms of non-violent resistance might we take? Who are we protesting against?

What arguably matters most with inheritance is what we do with it – the decisions we make in every given moment. Protest is the front line of resistance to our collective inheritance. History has been shaped by power and protest. Protest can arise in different ways: the hard-core, physical activism shown by people like Emma; softer attempts at political engagement and conversation; the courtroom fight to change laws; not to mention songs, art and culture as expression of a dissenting voice.

Protest is a long game

The reasonable man adapts himself to the world;
the unreasonable one persists in trying to adapt
the world to himself. Therefore, all progress
depends on the unreasonable man.
George Bernard Shaw, *Man and Superman*

In 1722 a man named Thomas Coram, moved by the frequent sight of abandoned babies dying in the streets of London, began a campaign to establish an orphanage.

After many years of campaigning, Coram managed to get the signatures of twenty-one aristocratic women to take to King George II, asking to set up the children's charity. When that petition failed, he asked the women to contact their husbands and male friends, to create a new larger petition of male signatures. After seventeen years of tireless campaigning, in 1739 the Foundling Hospital was established, offering the world's first incorporated charity. It was to operate until 1954, taking in and caring for approximately 25,000 children.

Protest is a long game. A formative moment of my childhood was 15 February 2003, when an estimated 1 million people marched in London against Prime Minister Tony Blair's decision to join President George W. Bush's call for war in Iraq. It was part of a wider movement of marches around the world: an estimated 6 million to 30 million people took to the streets, described by commentators as the 'largest protest event in human history'.

The fact that those marches were ignored, and that the US and UK still declared war on Iraq, felt like a defining moment of the political fabric of that era – as much as the terrorism that preceded and followed it. It was deeply disheartening, and disturbing, to see that millions of people could be so fundamentally ignored by the political establishment elected to represent them. It was a spiritual blow to recognize that however many turned out, it made no difference to the political outcome.

Yet in spite of that political deafness, in spite of the fact that we still went to war and unleashed a period of cruel, political instability with unimaginable suffering, that march did make a difference. It offers proof of the moral compass of our citizens at the time; that the war was not done in our name: that it was a failure of our politics, not our people. And as contemporary politicians now try to distance themselves from the decision to start that unpopular war, hopefully it will serve as a caution to future attempts.

The bitter-sweet reality is that protest takes time. It often does not overturn a decision first time around, but takes a sustained campaign of years – often decades – to build a sizeable movement to effect change. If you look back at many of the social victories we have made historically, such as the abolition of slavery, the right of women to vote, the civil rights movement, or the campaigns to drop 'developing' world debt, protest has been challenged and sustained long before victory. In the arena of environmentalism, protest has been a vital part of shifting public opinion for decades, and feels more potent now than ever, as it witnesses a wave of new and disparate voices rise up.

Giving voice to the voiceless

Protesting about new roads has become that
rarest of British phenomena, a truly populist
movement drawing supporters from all walks
of life.

The Economist, 1994

A smudgy black-and-white newspaper photograph of my cousin Emma, aged sixteen, dangling and chained to a tree, was the first image of environmental activism I had ever encountered. I was eight years old, a similar age to Emma when her own activist thoughts began, and she planted seeds that germinated in me over time.

Emma's teenage activism was fuelled by the road protests that swept the British public's imagination in the 1990s. In 1989 the Department of Transport had announced a £23-billion national road programme which boasted to be the largest of its kind since the Roman Empire, but would destroy many designated wildlife sites in its wake.

An unlikely assortment of people – environmentalists, locals, children, pensioners, conservatives and liberals – joined together at different sites across the country to generate opposition to the road expansion.

'I was so young and naive,' Emma reflects. 'The route of the bypass in Newbury, for example, went through huge swathes of ancient woodland, 10,000 trees to be felled on the route. I fell in love with the place and I was completely heartbroken when I watched them destroy it. This was a process that was repeated again and again over the next few years.'

But not every protest was lost. Developers had been given permission to move two rivers in South Devon, in order to extend their quarry. The local community had long been fighting the development plans without any success, scared of the risk of flooding and the damage to the local environment. Then about twenty 'tree-people' – including Emma – moved in, building tree-houses next to the river, and joined the locals on a march to London.

'These wonderful people arrived, and they really turned things round for us,' said villager Joy Coombe. 'They had the brains and the know-how, they knew who to lobby, and how to deal with the media.' An official public inquiry found that the developers' figures

were incorrect and rare species would be threatened by the development. Planning permission was
revoked. 'I will remember those days for the rest of
my life,' said Coombe. 'The eco-warriors showed that
you have to work really hard, and make sacrifices, to
get anywhere at all.'

Though other individual site protests were less
successful, the campaign of road protests generated
front-page news in the UK for years, shifting public
opinion and official policy. The British Department
of Transport ended up building only 37 out of the
original proposed 600 schemes. Steven Norris, the
transport minister who oversaw the Newbury bypass,
later admitted that 'the protestors were right' and the
government had been too motorist-centric. At the
end of five years of road protests, BBC reporter Paul
Clifton concluded, 'The protesters lost the battle. But
perhaps they won the war.'

My cousin Emma,
with Rosie in Aller-
combe, locked on
to tunnels to avoid
eviction. *Road
Protest* by Andrew
Testa, 1996.

Do not let your parents do anything that makes you dislike them

You are the bows from which your children as
living arrows are sent forth.
Khalil Gibran, *The Prophet*

In an essay published in the spring of 2018, for *Dark Mountain* journal, Emma reflected on our hypocrisy when it comes to consideration for the next generation: 'Society's worst outrage is reserved for the wrongs done to children. Yet, the wrongs that are being done to the world on a daily basis are the legacy we leave for them.'

A similar message was carried to world leaders at the UN Climate Change Conference COP24 a few months later by the fifteen-year-old climate activist Greta Thunberg: 'You say you love your children above all else, and yet you are stealing their future in front of their very eyes.'

Greta's activism came from learning about climate change in school, which caused her to become deeply depressed. Too young to vote, without money to invest and with a relatively small consumption impact (she is vegan and doesn't fly), Greta focused on her voice as a tool for change.

Her depression lifted as she first persuaded her mother to stop flying, and then persuaded her father – coincidentally named Svante after the scientist credited with recognizing human-induced climate change – to become vegan.

Svante Arrhenius usually gets the credit for discovering human-induced climate change, yet it was actually a female scientist (and women's rights campaigner) half a century earlier who theorized that changes in carbon dioxide levels could affect Earth's temperature. In 1856 Eunice Newton Foote ran a series of experiments and published a paper making this point. She was only recognized in 2011.

226

After the record heatwaves in northern Europe in 2018, which saw forest fires in Sweden all the way up to the Arctic, Greta began skipping school to sit outside Parliament in the run-up to the national elections, bearing a handmade sign that said in Swedish, 'School strike for climate'.

'I painted the sign on a piece of wood and, for the flyers, wrote down some facts I thought everyone should know. And then I took my bike to the Parliament and just sat there. The first day, I sat alone from about 8.30 a.m. to 3 p.m. – the regular school day. And then on the second day, people started joining me. After that, there were people there all the time.'

Greta continued striking every Friday thereafter, demanding that the Swedish government create climate policies in line with the Paris Climate Agreement. Her small, simple act of civil disobedience inadvertently sparked a global movement which within six months would see 1.4 million children across 128 countries strike school to demand climate action. Within a year an estimated 6 to 8 million adults and children had joined the associated Global Climate March.

Greta was diagnosed with Asperger's syndrome, OCD and selective mutism which, she says, 'Basically means, I only speak when I think it is necessary.' She has a terrific knack for the blunt.

Well-versed in climate science and policy, Greta quotes carbon budgets, IPCC reports and extinction rates to demand transformational climate action, and

When he wasn't writing about spending time on his own in nature, American essayist and philosopher Henry David Thoreau espoused the importance of being disobedient to an unjust government. He spent a night in jail for refusing to pay his poll tax in protest at the Mexican–American War and slavery (it is thought his aunt bailed him out). His arguments have inspired many activists since, including Martin Luther King, Gandhi, the women's suffrage movement and more recently Extinction Rebellion.

'new politics' that are in line with the demands of science, and climate justice.

Dream but stay woke

I've said it once, and I'll say it again, we can't eat
money, or drink oil.
Autumn Peltier, 28 September 2019, speech
at United Nations

Greta has become the poster child for the Youth Climate Movement, but there are other, less visible, youth activists who have been active around the world for years. Many of them understand the interlocking nature of injustice, and how environmental issues disproportionately affect marginalized groups. A study by the EPA, for example, has shown that people of colour are more likely (than white communities, or poor communities) to live in areas affected by environmental pollutants, indicating systemic environmental racism.

When eight-year-old Amariyanna 'Mari' Copeny wrote to President Obama about the water crisis in Flint, Michigan, she didn't expect a reply: 'My mom said chances are you will be too busy with more important things,' her letter read. At the time, Flint's 100,000 residents were drinking water so contaminated it led to twelve deaths and seventy-nine sicknesses. Luckily, her mom was wrong. Obama came to Flint to meet with Mari, and authorized $100 million to repair the water system.

Autumn Peltier from Wiikwemkoong First Nation on Manitoulin Island, Canada, has earned the name 'water warrior' after fighting for water conservation and indigenous water rights since she was eight. In Brazil, Artemisa Xakriabá, a teenager from the Xakriabá peoples, describes how mining companies have denied her community access to the river and campaigns to stop environmental destruction in the rainforest. After the Amazon fires brought international attention to the devastation, in 2019, she told Western journalists: 'The main thing you can do in the West to help is to stop importing hardwood.'

Isra Hirsi, who co-founded the US Youth Climate Movement when she was sixteen, says she came to environmentalism 'late' after first engaging with the Black Lives Matter movement. At first, she considered climate change a 'white issue' but then realized it is also fundamentally an issue of justice.

The sun also rises

I think there's a weapon of cynicism to say, 'Protest doesn't work. Organizing doesn't work. You'all a bunch of hippies.' Frankly, it's said out of fear, because [protest] is a potent force for political change.
Alexandria Ocasio-Cortez, *Democracy Now!*, interview with Amy Goodman

Morissa Zuckerman set up the Bay Area chapter of the US youth climate Sunrise Movement after her experience working on a fossil fuel divestment campaign at her college. 'Organizing made me understand the power of people,' she tells me in a phone call, 'of what we can do together.'

The divestment movement began on US college campuses, targeting universities investing in fossil fuel extraction. When I was at Cambridge University in 2009, I joined an economic board to voice the possibility of divesting. To a room full of old white men sitting on high-backed chairs, I tried questioning investments into the arms trade and fossil fuels, but was met only with hard, cold stares. Ten years later Cambridge are seriously considering divestment from fossil fuels.

Meanwhile, in 2019, 'Cambridge Zero' was launched: an initiative to use the university's vast interdisciplinary research power to help support the transition to a zero carbon economy. The university has committed to have 'absolute zero' net carbon by 2048.

Meanwhile, other university campaigns have had quicker success. In 2014, Glasgow University committed to divesting from fossil fuels, and Stanford divested its endowment ($18.7 billion) from coal extraction after an undergraduate campaign, in a move that Naomi Klein called 'the biggest victory to date'.

Yet, in spite of their successes, Morissa says, 'We realized the fossil fuel divestment campaigns weren't going to go far enough to match the scale of the problem. Which led to the founding of the Sunrise Movement.'

In November 2018 over 250 Sunrise Movement protestors occupied the office of Nancy Pelosi, Speaker of the House of Representatives. They demanded that all Democrats in the House refuse donations from fossil fuel companies, and that Pelosi establish a 'Committee on a Green New Deal' to work on climate legislation. Fifty-one of the activists were arrested, and Pelosi offered to reinstate the 'Committee on the Climate Crisis'.

A newly elected Representative, Alexandria Ocasio-Cortez (or AOC as she is colloquially known), attended the Pelosi office protest and was inspired. Formerly a bartender, the self-professed 'Puerto Rican girl from the Bronx' had entered politics sideways. Her brother had nominated her to 'Justice for Democrats': an NGO seeking to find and support diverse political candidates who would eschew corporate money and disrupt the system from inside. AOC usurped the ten-time incumbent in the Democratic primaries, won the general election, and became, aged twenty-nine, the youngest woman ever to serve in the United States Congress.

The first thing AOC did as a member of Congress was work with the Sunrise Movement to propose a Green New Deal: an ambitious outline for national climate action that has made waves nationally and internationally, raised the bar on environmental policy, and put climate change at the forefront of US presidential campaigns for 2020.

'It's been incredible to see the way this movement has changed the baseline of what we can expect from policy,' Morissa says.

Is she optimistic? I ask.

'My mom always tells me the saying, "Pessimism of the intellect, optimism of the will." Action really is the best antidote to despair.'

Counter-lobbying

> The real power belongs to the people.
> Greta Thunberg, 12 December 2018,
> speech at United Nations

After Greta joined 20,000 children in Paris on a school strike, she and several other young campaigners met with President Macron. When I spoke with Macron's advisor Aurélien Lechevallier, he blamed the slow political response to the climate crisis on a 'disconnect' between international agreements, grass-roots activism and national policy; and the insidious impact of lobbying.

'There is always a balance in the political priorities. We have so many local protests and we see great speeches and conferences, but the pressure of the lobbies of the fossil fuel companies and the financial system is really too strong to change the way the national policies work. Lobbying is very difficult to find and to track. It's around all the decision-making bodies. It's a constant influence on different leaders and congressmen and journalists. Every day is a suffering because we know that we are moving too slowly,' Lechevallier said.

The world's five biggest publicly listed oil and gas companies, including BP and Shell, spend nearly $200 million a year lobbying against environmental regulation. Yet Lechevallier is optimistic that the protest movements will act as a counter-force to this: 'Now with the rising of public voices, I think it's becoming more difficult for the lobbies to deliver their messages.'

Indeed, Mohammed Barkindo, secretary general of the trillion-dollar Organization of the Petroleum Exporting Countries (OPEC), called the Youth Climate Movement 'the greatest threat' to the fossil fuel industry. He said that a 'mobilization' of world opinion against oil was 'beginning to dictate policies and corporate decisions' and that the organization's children are 'asking us about their future because . . . they see their peers on the streets campaigning against this industry'.

If you love the law

An individual who breaks a law that conscience tells him is unjust, and who willingly accepts the penalty of imprisonment in order to arouse the conscience of the community over its injustice, is in reality expressing the very highest respect for law.
Dr Martin Luther King Jr, 'Letter from
a Birmingham Jail', 1963

Does civil disobedience imply disrespect, or the highest respect, for the rule of law? Farhana Yamin has gone full circle. She began her journey as an environmental lawyer, optimistic that she could use the law to guide progressive environmental change. She worked on the EU's pioneering emissions trading scheme,

co-authored several IPCC reports, and helped nego-
tiate the Paris 2015 Climate Agreement.

Then, in 2016, knowing the Paris Agreement
wasn't ambitious enough, watching the US withdraw
from it, and observing the overall failure of the legal
system to rein in global emissions, she fell into a deep
depression.

'I felt like a failure,' she told me. 'I'm this lawyer
who's worked her whole life, leaving my kids for two
or three months at a time to go through all these international processes, but the hopes that I had for law
and for multilateralism, had all kind of evaporated. I
felt like we'd been duped and I'd duped myself, thinking that law was going to fix it all.'

Farhana stopped doing her environmental work
until the IPCC issued its 2018 report: the result of
nine years of work which she had originally been involved with, warning that the world faces climate
catastrophe if all systems are not changed by 2030. 'If
you're alive and a sensitive human being, the report
makes you cry,' Farhana says. 'It makes me cry.'

> Scientists might want to write in capital
> letters, 'ACT NOW, IDIOTS', but they
> need to say that with facts and numbers.
> And they have.
> Kaisa Kosonen, Greenpeace's response
> to the 2018 IPCC report

Farhana watched as various NGOs she worked with
issued 'really beautiful press releases', but a small

group of activists responded in a more surprising way: by staging a 'rebellion'.

On 31 October 2018, more than a thousand people took to London's Parliament Square to see the 'Declaration of Rebellion' against the UK government. Greta Thunberg, Caroline Lucas and George Monbiot read speeches, standing amongst the statues of Millicent Fawcett, Gandhi and Nelson Mandela – who in their time all partook in civil disobedience. The activists staged a 'sit-in' in the road, and fifteen were arrested. They called them themselves Extinction Rebellion (XR).

'Something in my heart was like, that's the right response,' says Farhana. 'It's time to stop issuing press releases: it's time to put our bodies and our liberty on the line.'

Beyond protest

Courage calls to courage everywhere.
Millicent Fawcett

Since 2018 XR have grown in scale and ambition, in hundreds of cities worldwide. Their manifesto calls on governments to 'tell the truth' and declare a climate emergency; commit to carbon neutrality by 2025; and set up Citizens' Assemblies to help oversee the transition.

XR are committed to peaceful non-violent resistance, and seek to cause economic impact by blocking bridges and roads, disrupting government buildings, doing <u>hunger strikes</u>, and practising tax resistance.

'XR is not a protest,' says co-founder Dr Gail Bradbrook. 'It is not a campaign. It is a rebellion. It's a shift in the consciousness of each of us. We are in active rebellion against our government. The social contract is broken, the governments aren't protecting us and it's down to us now.'

In November 2018 the first wave of large XR protests saw five major London bridges blockaded by protestors. The action, which included over 6,000 people of whom 70 were arrested, was described by the *Guardian* as 'one of the biggest acts of peaceful civil disobedience in the UK in decades'.

It was indeed a peaceful atmosphere. The dedication of the activists, and the sombre mood as they listed species lost already – or close to extinction – made me tearful. A young woman called Blythe stood on a concrete slab being heckled by police. 'I'm just going to finish this,' she politely told the officers, resisting arrest long enough to get to the end of her speech, in which she explained she had chosen not to have children because the future feels too painfully uncertain.

Police had been warned of the event in advance and so offered a palpable presence on the bridge. I asked one of the police officers how she felt about the event: was it frustrating? 'I don't agree with them, but they are exercising their right to peaceful protest,' she replied, almost warmly.

In 2019, I visited two grandfathers (Peter and Marko) who were on hunger strike for over three weeks, in an attempt to speak with UK Government representatives about the climate crisis. They didn't get a meeting. Peter told me they were making this sacrifice so they could 'look our grandchildren in the eye'.

There is no time to delay changing course
radically to avert major catastrophe. The
activists of Extinction Rebellion are leading
the way in confronting this immense challenge,
with courage and integrity, an achievement
of historic significance that must be amplified
with urgency.
Noam Chomsky, *Guardian*, 'Climate
activists and police tussle for control of
Oxford Circus'

In 2019, XR launched multiple international campaigns, blocking roads and disrupting transport for
weeks at a time, with humour and creativity: staging
music, speeches, yoga classes and even mass breastfeeding sessions in prominent sites in cities.

XR's protestors often actively incite arrest, and
many have been made – including that of Farhana
after she glued herself to the pavement in front of
Shell's offices.

'This is not lawlessness, because we're choosing to
break specific laws, peacefully and consciously, and
submit to the consequences, in order to make the
point that the laws themselves are not just, and require a systemic overhaul,' she said.

'I thought that the law and science and speaking
truth to power worked. That we would be able to act
with kindness and in time, [but] we haven't done it.
That outcome is not some random delay; it has been
planned for, financed and lobbied for. That's why I
feel totally comfortable protesting.'

The right to a free conscience is enshrined in international law. In 2008 a jury in the UK ruled in favour of a group of Greenpeace activists who had scaled the pipe of a coal plant to spray-paint it. They accepted the activists' argument that the damage they had caused to the coal plant was much less than the damage that was being caused by coal in relation to climate change.

'I'm prepared to go to jail,' Farhana told me. 'I'm prepared to go to jail repeatedly. [But] if the jury understand the gravity of the crisis and are willing to accept that argument, that shows we've won the hearts and minds of ordinary people.'

Is it popular?

When Gail Bradbrook first starting asking people in 2010 if they would be willing to get arrested, 'You might as well have asked someone to get naked and take a shit, in the room,' she reflected, years later, when the arrests were in the thousands.

'It's often the case when you talk about changing things, that people think you're talking about somebody else . . . I actually think, it's about all of us. We often think that the power is in the media, or Westminster, but the power lies in the collective. What are we actually willing to do? Some of us have to be willing to die for this thing.'

A few weeks after the bridge protests, I met with Gail in a London train station for a coffee (in her reusable cup). She seemed gentle, and told me in her

Yorkshire accent about research which suggests that forms of non-violent resistance have historically been twice as likely to succeed as violent resistance, and that 'no campaigns failed once they'd achieved the active and sustained participation of just 3.5 per cent of the population'.

Is the desire to protect the planet popular enough?

XR's tactics have been polarizing. One British poll found that 60 per cent agreed with XR's aims, but 53 per cent disapproved of their tactics. They have been called too white, too radical, too hypocritical. Yet, despite – or perhaps because of – the thorny, polemical response to XR, and the sacrifices they are willing to make, they have successfully generated unprecedented debate and media coverage, getting the climate crisis on to the political agenda.

Their meteoric rise coincided with the School Strikes, the work of David Attenborough on plastic pollution, unusual weather, scientific reports and a groundswell in public consciousness on the issue of climate change. By 2019 two thirds of Britons said they wanted their MPs to support ambitious new environmental plans and laws; whilst 60 per cent of Americans identified as 'alarmed' or 'concerned' about global warming.

The political landscape has shifted.

In 2019, the Green Party made significant gains across the UK in local elections and swept European elections with a record number of MEPs.

Within a few weeks of Farhana's arrest, the UK government ceded to one of XR's demands, and became the first country in the world to declare a climate emergency, swiftly followed by other nations and hundreds of local governments around the world.

Before resigning as Prime Minister, Theresa May's swansong was to commit the UK to a legally binding target of net zero carbon emissions by 2050, becoming the first G7 country in the world to do so. The House of Commons Select Committees called for Citizens' Assemblies to help define a pathway to meet the goal of net zero: fulfilling the third of XR's three requests.

It turns out if you *really* respect the law, you sometimes have to break it.

New laws for planet Earth

Exercising my 'reasoned judgment,' I have no
doubt that the right to a climate system capable
of sustaining human life is fundamental to a free
and ordered society.
US District Judge Ann Aiken, Juliana v.
United States
Youth Climate Lawsuit

Instead of fighting against the law in acts of civil disobedience, and risking arrest, some activists seek to *use* the law, to try to *change* it. Children have been at the forefront of numerous legal challenges against companies and governments, to better protect the environment.

In 2015, twenty-one young people – aged between ten and twenty-one – filed a lawsuit, 'Juliana v. United States', claiming the US government had knowingly failed to protect its citizens from climate change, and seeking to end dependence on fossil fuels. The historic case received support from members of the US Congress, legal scholars, historians, doctors, lawyers, businesses and more than 32,000 young people, and looks set to proceed to trial at the US Supreme Court, despite efforts by both the Obama and Trump administrations to have it dismissed.

One of the children behind the case is the granddaughter of the NASA scientist Dr James Hansen who in 1988 first warned Congress of climate change. He testified on behalf of the children, saying, 'In my opinion, this lawsuit is made necessary by the at-best schizophrenic, if not suicidal nature of US climate and energy policy.'

The case is backed by the non-profit Our Children's Trust, who are supporting similar legal cases in different states and around the world, whilst other lawsuits have been launched against the EU by children and families, for climate change damages.

In 2015 the world's first climate liability suit was won: the District Court of The Hague ruled that the Netherlands government must do more to reduce greenhouse gas emissions, after the case was brought forward by 900 people against the government two years earlier. The ruling set a precedent, with Client Earth chief executive James Thornton commenting that the 'reasoning is applicable in any legal system and will certainly be used by courts in other countries'. Similar cases have since been made by community groups in Belgium, Norway, the United States, South Africa, India, Switzerland, New Zealand and Pakistan.

Lawsuits are also being directed at companies. In 2009, a coalition of NGOs and cities won a lawsuit brought against a US state-owned bank, accused of financing fossil fuel projects in violation of National Environmental Policy. In 2017 and 2018 several US cities, and a consortium of fishermen, launched lawsuits against multiple fossil fuel companies for damages due to climate change and rising sea levels.

'Some of the most pioneering litigation is coming out of developing countries,' Farhana told me, 'which is absolutely brilliant because that's where all the oil speculators live.'

Ecocide

> Ecocide could make the difference between a habitable and an uninhabitable planet.
> George Monbiot, *Guardian*, 'The destruction of the Earth is a crime. It should be prosecuted'

'You might say that the Earth is in need of a good lawyer,' Polly Higgins told her parents before embarking on a career in law. Yet like Farhana, Polly reached the conclusion that litigation was not enough. Instead, she quit her job, sold her home, and spent the last ten years of her life trying to make a 'missing law' – a law that criminalized collective harm.

'In the first RBS meeting, there was a press conference and the CEO was asked, "Why are you financing the exploitation of the Athabasca tar sands?" And he just laughed and said, "It's not a crime!"' Polly recounted, 'With an ecocide law in effect, corporate offences could be prosecuted, with directors and officers held personally liable; governments would not be able to give out licences or ministers could be prosecuted; finance would not be forthcoming.'

Efforts to establish ecocide as a crime date back to the 1970s when the Rome Statute was first drafted. The statute, overseen by the International Criminal Court, applies its laws – including 'genocide', 'crimes against humanity', 'war crimes' and 'the crime of aggression' – globally, over and above domestic rules of law. Ecocide was included in the original draft but removed at the last minute because of corporate lobbying in the four objecting countries: the US, UK, Netherlands and France.

In 2009, Polly revived the attempt to establish ecocide as a crime. She lobbied the UN, wrote a book, organized mock trials and set up a trust where individuals can donate and become signatories to the idea. People who sign up are called 'Earth Protectors'. I am one.

'Destroying the Earth is a sin,' said Pope Francis, who has voiced public support for ecocide to be made an international crime.

Polly remained optimistic because to amend the Rome Statute requires only one head of state – out of 124 – to call for an amendment, then an assembly would be convened within three months. If the majority of member states agreed, no individual country could veto it. Polly began working with nations already struggling to cope with extreme weather, such as the Polynesian island state of Vanuatu.

Vanuatu's foreign minister, Ralph Regenvanu, said, 'My government is now exploring all avenues to utilize the judicial system in various jurisdictions – including under international law – to shift the costs of climate protection back onto the fossil fuel companies, the financial institutions and the governments that actively and knowingly created this existential threat to my country.'

In March 2019 Polly was diagnosed with an aggressive form of cancer and given only weeks to live: 'There are millions who care so much and feel so powerless about the future, and I would love to see them begin to understand the power of this one, simple law to protect the Earth – to realize it's possible, even straightforward,' she wrote to George Monbiot. 'I wish I could live to see a million Earth Protectors standing for it – because I believe they will.'

Sudden media spotlight was thrown on her proposal for ecocide and tributes poured in from around the world. A few weeks later Polly died, but the fight for ecocide didn't. 'Her work will live on,' tweeted Naomi Klein, whilst XR activists pledged to continue her fight, spray-painting on the wall of a fossil fuel corporate during their protest: 'FOR POLLY ♥'.

The possibilities of protest

Protest is a privilege. Already in some countries around the world protestors are being <u>met by police</u> with violent force, tear gas and pepper spray. Meanwhile, an Environmental Defender continues to be killed every one to two days, with little recourse to justice.

'I am privileged,' acknowledges Farhana. 'I live in a safe country, where protest is possible.' In the UK, the ability to be arrested is also a social privilege: 'Not everyone has to go to prison: most people will not be able to take time off, many have a lot of responsibilities, or wouldn't be treated fairly in this system.'

> Non cooperation with evil is as much a moral
> obligation as is cooperation with good.
> Dr Martin Luther King Jr,
> *The Autobiography of Martin Luther King, Jr.*

In the UK, counter-terrorism police briefly listed XR as an extremist group. If you've been to an XR event, 'terror' is hardly the appropriate way to describe the look of passers-by: the spectrum probably ranges from 'warm interest' to 'annoyance'. Police later recalled the document, saying it was an error of judgement.

Many people don't feel comfortable marching or risking arrest. Meanwhile, the rise of the Internet has enabled movements to erupt overnight in disparate ways. On the one hand you can criticize online activism as 'armchair politics' that breeds apathy, yet on the other hand, the Internet is enabling a public dialogue, and making protest accessible for people who would, or could, not otherwise take more physical, direct action.

My mother is the perfect example of this, and the source of many of the campaigning emails I receive each week. She is someone who cares deeply, but is too disabled to attend most events. She came to the Global Climate Strike with me and my daughter, in a wheelchair.

In reality, protest takes numerous forms. You don't need to march or glue your arse to a concrete slab. There are many nuanced ways to engage and be heard: through journalism, supportive comments, donations, conversation, consumer boycotts, petitions, speaking to representatives; also through art, music or subtle non-cooperation with the status quo.

Arguably the most important mode of protest is that which is carried home and into our daily lives, beliefs and small choices. The recognition that the system isn't some evil force outside of us, but a product of our collective acts and decision making.

The wisdom to take responsibility and protest against aspects of our own selves.

The courage to really, truly, care.

Caring about the future

Of all the dangers we face, from climate chaos to nuclear war, none is so great as the deadening of our response.

Joanna Macy, www.joannamacy.net

My daughter and her three-year-old friend played in the seats as XR co-founder Tamsin Omond gave a speech on extinction and the need for transformation. They giggled and laughed, Wylde repeating random words from the speech.

It was a strange and poignant moment.

In the three years I have been writing this book, this whole chapter – XR, the School Strikes, the Sunrise Movement – has emerged. Environmental protestors are at the front line, but their sentiment runs wide and deep.

When enough people care, politics acts.

'Always, I have pushed to change the future, no matter how small my impact,' says my cousin Emma. 'The future has pushed back, but it's not a battle. We are inseparable. I will never stop trying to shape it, change it, dent it. Pushing for positive change, even against the odds, is part of my sense of purpose. It makes me feel more in control, in a system which is so out of control.'

Beyond defending voiceless species, environmentalism is a new civil rights movement, intersecting with race, gender and poverty: for the rights of vulnerable and marginalized communities around the world today; and for those – like my daughter and her friend – too young to vote; and for their children not yet born. Fighting for the future.

Chapter Nine
A Pacifist's War

Net zero, carbon pricing and upgrading capitalism

In 2006, the sardonic and popular animation *South Park* made an episode that offered a love letter to climate-change denial. A cartoon Al Gore was hysterically warning everyone of a monster called <u>ManBearPig</u> – 'half man, half bear, half pig', representative of global warming. In the process of trying to find it, the characters met calamity after calamity; in the end ManBearPig was deemed a dangerous and non-existent ruse to get Al Gore more attention.

In 2018, *South Park* issued an atypical hour-long apology, 'self-flagellating' themselves. Al Gore returned for a new episode in which ManBearPig was deemed real. The characters begged Al to help them track it down.

'Say you're sorry,' said a bruised Al, before agreeing to help.

Climate change has gone from being a niche concern, decades away – a pop joke, and online meme – to an imminent political priority.

Responding to this swell in public consciousness, and the cascade of doomsday scientific reports, there has been a rise of diverse politicians proposing ambitious ways to address the crisis: and even to turn it into an opportunity.

Turns out folks in *South Park* have been making deals with ManBearPig, to come back at a later date, for generations past. The villagers are told they can now get rid of ManBearPig forever – all they have to do is sacrifice soy sauce and their favourite video game. Which, of course, they don't want to do – making a new deal with ManBearPig to come back in five years instead.

The scale of the task at hand for any politician who accepts the demands of science is not to be underestimated: huge sunk costs; widespread infrastructure to replace; whole industries to shift. Alongside the realities of adaptation, an increasing number of climate refugees and the escalating costs and social toll of extreme weather.

It's no wonder some want to live in denial. Wouldn't it be great if global warming was a hoax, and we could just kick back and stop worrying? Yet, physics doesn't debate or delay. It's high time to put science before politics.

A pacifist's war on warming

Our destination is clear: deep decarbonization, delivered as humanely as possible. According to the IPCC the whole world must be carbon neutral by 2050, to avoid warming over 1.5°C, and risk losing all global coral reefs, displacing millions – if not billions – of people and potentially triggering <u>runaway climate change</u>.

Two countries – Bhutan and Suriname – have already achieved carbon neutrality, and seventy-seven countries and over a hundred cities and states, at the time of writing, had agreed to be net zero by 2050, including sizeable emitters like France, Germany, the UK, California and New York. Most European countries have agreed to get the whole EU to commit to

Not sure what runaway climate change is? Check out the appendix at the back of this book for some context.

be carbon neutral by 2050: increasing climate-change funding to help get there.

All commitments to net zero by 2050 are cause for celebration.

But then we should pause and reflect on the fact that it's not ambitious enough.

The year 2050 largely defers responsibility not just to a later generation but to later governments, which can contradict short-term policy: the UK increased fracking and expanded Heathrow airport whilst announcing their goal; Canada approved a new oil pipeline the day after their declaration of a climate emergency.

It's also risky.

Would you wait until the very last minute?

And if wealthy countries are waiting to the very last minute, how can we expect poorer countries to speed ahead?

Everyone must get to net zero by 2050, or we all go down.

To this end, Steve Trent from the EJF recommends aligning development policy and aid budgets with climate strategy. As an example, he says, 'If a substantially larger proportion of aid budgets was applied to solar energy, this would provide life-changing savings on fuel costs, particularly for the poorest while reducing the respiratory problems and burns from kerosene.'

The aid budget to sub-Saharan Africa, for example, is €9 billion, which if diverted to renewable energy, could help people become energy independent on solar, and resolve many of the problems (and hospital admissions) highlighted by Steve Trent.

After the UK and Germany announced 2050 as a goal, the environmental protests continued, asking for greater ambition. Multiple studies also show that total decarbonization (in countries like the UK) is possible within ten years. What is required, many argue, is a warlike response, that sees immediate shifts in policy to have us well *on track* to net zero by 2030.

In the language of environmental political action, war references abound. Climate change is often called the Third World War. In 2014, China's Premier Li Keqiang announced a 'war on pollution'. In 2018, Alexandria Ocasio-Cortez (AOC) said, 'The Green New Deal we are proposing will be similar in scale to the mobilization efforts seen in World War Two or the Marshall Plan.'

Although it can be unnerving to pacifists like myself, war offers us a helpful metaphor, as it captures the scale of the task, and reminds us of our capacity to work together to make enormous transitions very quickly. During the Second World War, within the space of five years agricultural production in the UK doubled, and 2 million women were brought into the workforce. In both small and large ways, everyone participated in the effort: from giving their lives to turning off the lights at night.

In an increasingly divided world, a war on warming also offers the chance for us to unite across political,

Rationing food, raising taxes and car-pooling were some of the collective home acts encouraged during the Second World War. In 1942, US President Roosevelt told millions of US citizens: 'When at the end of this great struggle, we shall have saved our free way of life, we shall have made no "sacrifice".'

In *The More Beautiful World Our Hearts Know is Possible*, Charles Eisenstein conversely resists using the mainstream narrative of War on Climate Change, as it 'subordinates all the small, local things we need to do to create a more beautiful world to a single cause for which all else must be sacrificed'.

economic and national boundaries, against a common threat that will ultimately impact everyone.

A Green New Deal

Green is not about cutting back. It's about creating a new cornucopia of abundance for the next generation by inventing a whole new industry.
Thomas Friedman, *New York Times*, 'The power of green'

The concept of a 'Green New Deal' was first posited in 2007 by Pulitzer Prize-winning author Thomas Friedman. It was a vision of how the government could leverage 'free-market capitalism' to mobilize environmental action on a scale comparable to President Franklin Roosevelt's 1933–36 'New Deal', which helped lift the United States out of the Great Depression through spending on infrastructure, public works and new regulations.

'The New Deal was not built on a magic bullet, but on a broad range of programs and industrial projects to revitalize America. Ditto for an energy New Deal,' Friedman wrote. 'We will only green the world when we change the very nature of the electricity grid. And that is a huge industrial project.'

Presidential candidate Barack Obama added a Green New Deal to his policy platform in 2008 (although it was later revised), and in 2009, the UN called for a Global Green New Deal, 'in response to

252

the multiple global crises of 2008 – fuel, food and financial'. The concept of a Green New Deal has long been advocated by a series of progressive politicians, including the UK and Australian Green Parties.

In February 2019, the Green New Deal (GND) regained international media attention, when US Democratic candidate AOC, and Senator Ed Markey released a resolution for a 'Green New Deal' which would aspire to shift the US economy to 100 per cent renewable energy within twelve years, retro-fit all buildings across the US to improve energy efficiency, and invest in cleaner public transport.

The Bill has received backing from the Sunrise Movement, over 600 other organizations, and many Senators and Democratic presidential candidate hopefuls. Even if it doesn't get passed, it has raised environmental policy ambition, and inspired similar policy proposals around the world, such as the Labour for a Green New Deal Group in the UK.

A Green New Deal had first been politically proposed in the UK in 2008 by Green Party MP Caroline Lucas. Revisiting the proposal ten years later, with renewed enthusiasm, she told me, 'It's immensely exciting to see the idea get such traction in the US. Politics is so important because it sets the long-term framework for action, and so much of what now needs to happen requires huge government investment: whether that's insulating homes, building renewable energy, or sustainable public transport. Behaviour change is important, but what we really need is systems change.'

AOC's vision of GND was not without its critics, even enemies, because it sought to package a carbon revolution with seemingly socialist policies: such as a job guarantee for all Americans, and universal basic healthcare.

A Republican think tank estimated it would cost between $51 trillion and $93 trillion over the next decade. US Green Party leader Jill Stein claimed her alternative proposal for a GND would cost $700 billion to $1 trillion annually, and Democratic Senator Dianne Feinstein said bluntly, 'There is no way to pay for it.'

'The cost of pursuing a GND will be far less than the cost of not passing it,' AOC countered, in an impassioned speech. 'This is not an elitist issue. This should not be a partisan issue. This is about our lives. We are facing a national crisis and if we do not ascend to that crisis – if we do not ascend to the levels in which we were threatened at the Great Depression, we were threatened in World War Two – then I don't know what we're here doing.'

AOC advocated raising the top marginal tax rate to 70 per cent in order to fund the programme, and in 2019 a group of billionaires in the US called on the government to introduce a 'wealth tax' to help deal with climate change and inequality. This has precedent: in the US during the Second World War the top tax rate was raised to 94 per cent.

In her book *This Changes Everything*, Naomi Klein makes the case that $2 trillion can be generated annually through a mix of largely popular measures, including closing tax havens and phasing out fossil fuel subsidies. Taxes could also be applied to luxury goods, frequent flights, and the robots increasingly displacing jobs.

'I hear people talking the language of participation and justice and equality and transparency, but no one raises the real issue of tax avoidance. It feels like I'm at a firefighters' conference and no one is allowed to speak about water,' Dutch historian Rutger Bregman said in a panel talk at the World Economic Forum in Davos, which went viral online.

'What must industry do to stop a broad social backlash? The answer is very simple — just stop talking about philanthropy and start talking about taxes. All the rest is bullshit in my opinion.'

At the other end of the spectrum is a GND led by business, rather than big government. 'The only thing as powerful as Mother Nature, is Father Greed,' Thomas Friedman likes to say. 'Forget the Space Race. We need an Earth Race: a free-market competition to ensure that mankind can continue to thrive on Earth.'

Research suggests that with the right regulations in place the private sector may even help fund a GND: sustainable investing is now a $12 trillion market in the US.

Though we might endlessly (and ideologically) debate the details of any proposed GND, and the roles that government or business should play, what is more important is that we make an ambitious and immediate commitment to reach net zero *before* 2050.

As Friedman says, 'Let a hundred Green New Deal ideas bloom! Let's see what sticks and what falls by the wayside . . . we don't have another decade to waste.'

Empowering change

> It is difficult to get a man to understand
> something when his salary depends upon his
> not understanding it.
> Upton Sinclair, *I, Candidate for Governor:*
> *And How I Got Licked*

One of the biggest political and psychological obstacles to mobilizing for the environment is the reality of change it demands – disrupting whole industries and livelihoods: 'It's difficult now because we are in this transition, and the standoff for the people is the old world with jobs and ways of producing and consuming that is not sustainable any longer, and they don't see exactly what is their role, their place in the new world that we want for our children, that is carbon free,' Aurélien Lechevallier told me.

I spoke with my mother about our family's experience of coal mining in South Wales. In spite of all of the implicit health impacts and dangers, the area really suffered when the mines suddenly shut down. 'The community had built up around the mines,' she said. 'When the collieries went, the area became very depressed.'

Likewise, on the other side of the world, Lily Gardner, fifteen-year-old member of the Sunrise Movement, grew up watching poverty grow, as people's jobs in the coal mines of Eastern Kentucky were displaced. Gardner poignantly observes, 'People are not climate deniers because they don't believe in facts, people are climate deniers because they're so afraid that they cannot confront another thing that is going to put them deeper into poverty.'

It is not surprising then, that much of the rhetoric around political action on climate change is focused on jobs. AOC's Green New Deal would see 'the creation of millions of high-wage jobs'; Labour for a GND intend to 'create good green jobs in every town and city across the UK'. Meanwhile critiques of the GND have often centred their attacks on the claim that it would 'destroy' millions of jobs, in older industries.

There is no doubt that re-skilling, and sensitive social policies, will be essential to meet the challenge of change, and companies and governments alike have a role in the process. For example, when the last coal mine in Centralia, Washington, was closed, a $55-million fund was created to provide job training and education to the families who had worked in the mine. Universal Basic Income is another policy we will look at in Chapter Twelve, which could help.

There is also an opportunity to transition equitable skills from old, polluting or destructive industries, towards new, green ones. The Campaign Against the Arms Trade (CAAT) suggests that the skills of the workers, and the factories themselves where armaments are made in the UK, could easily build renewable energy technology instead. This would position the UK as a world leader in an important, growing industry, rather than selling arms indiscriminately, in trade deals that the UK courts have deemed an illegal breach of humanitarian law.

An honest market

> The only reason we're not moving like gang busters to cleaner energy is because every system and policy in place subsidizes the dirty stuff.
>
> Lisa Jackson, Apple

Consider that three quarters of greenhouse gas emissions are caused by fossil fuels.

Consider that international agreements, decades old, have committed to limit greenhouse gas emissions.

Consider that our governments still spend over $5 trillion a year subsidizing fossil fuels.

Something does not add up.

The International Monetary Fund (IMF) calculated that governments are spending over $5 trillion a year by underpricing energy (85 per cent of which goes to coal and petroleum), and not factoring in their environmental and social costs, which will be billed to taxpayers now and in the future.

How to solve this?

One policy idea came up over and over again in the course of writing this book, from diverse perspectives and world views: making companies pay to pollute.

Economist Guy Standing maintains that anyone who pollutes the air should be made to pay, via 'ecological taxes' which would be redistributed to everyone equally as a 'carbon dividend': 'The air is part of our commons. Take air pollution and greenhouse gas emissions . . . if you allow the encroachment on our air, we should demand that the commoners – all of us – should be compensated.'

This idea – charging for pollution – is sometimes called a carbon tax, but it is not a tax. It is a *price*, that makes visible the very real but hidden costs of pollution and environmental damage. A society impacted by pollution and global warming cannot avoid paying these costs, one way or another.

When managed as a 'dividend', it is also a price that can circulate back to the people.

'If I would have a magic wand,' Lisa Jackson told me, 'it would make people see the true cause of the pollution that's around them   that cost of their sickness, the cost of their lost work, the cost to their kids, all those costs are borne by society. It doesn't have to be a tax – when people see what pollution really costs them, they make their smart choice.'

A price on pollution (such as carbon) would incentivize people and companies to minimize their environmental impact, improve energy efficiency, and ultimately flip the logic for consumers so that the more environmentally friendly products and services would ultimately become the cheaper ones (which they should be, because they are in reality).

It would make the market more honest; make it easier for social businesses to succeed, and for consumers to better navigate the impacts of their shopping choices.

The lack of a carbon price in most countries 'causes the economy to not function properly, and to over-pollute', says Elon Musk. 'The price of carbon is not zero. There is a real cost to it. There are things that are done to subsidize electric vehicles and subsidize solar, but those are not nearly as effective as simply putting a price on carbon.'

A carbon price also offers us the opportunity to create a fund worth hundreds of billions. What might a policy like this afford?

New possibilities

People need to feel the benefits of the carbon price in some way. A 2018 US survey found that 67 per cent of respondents would support a carbon tax on fuels if the funds from the tax were invested into the restoration of forests, wetlands and other natural features. This could go some way towards the vision of re-wilding large areas of land that we looked at in Chapter Seven, as a major climate solution.

Looking at the examples of when carbon pricing has worked well, communities and low-income individuals have been placed front of mind to alleviate the adverse effects if consumer product prices go up as a consequence. British Columbia's pilot of a carbon price in 2008, with wealth redistribution built in, reduced carbon emissions at ten times the rate of the rest of Canada, and had majority support in polls.

The British Columbia pilot was designed so that everybody receives an equal amount of money back as a carbon dividend: meaning that people who pollute less end up benefiting. It was so successful that in 2019, Prime Minister Justin Trudeau mirrored it nationally, across Canada.

A price on carbon has advocates across the political spectrum. US Republican Representative Bob Inglis suggested making a carbon price revenue neutral, offsetting the income with reductions in other taxes: 'It's essential for us as conservatives that action on climate change does not result in the growth of government.' He calls it a '100 per cent returnable emissions tax' because 'when you get to carbon, a lot of my fellow conservatives break out in hives, and then go into anaphylactic shock when you say the word "tax"'.

Inglis argues that the US could show 'bold leadership' and pave the way to an international price on carbon, by making a domestic carbon tax border adjustable. He suggests: 'The United States steps in and says, "We just priced carbon dioxide and if you're shipping stuff in here, we're collecting on entry. But if you do the same thing in your country, great, you won't pay any landing fees here."'

This concept was familiar from the ideas of Yale professor William Nordhaus, who shared the Nobel Prize for economics in 2018 for his work on carbon pricing. Since the 1970s, Nordhaus has advocated a carbon price as a more efficient and effective way to lower carbon emissions than direct government controls. He proposes making a 'global climate club' to deal with the international nature of the problem: the club would have a critical mass of countries who agree to a climate price, then could impose tariffs on non-members (in effect encouraging them to join the club).

There are other ideas about how to use the revenue from carbon pricing. Friends of the Earth recommends investing in home insulation, which would have the double benefit of reducing energy use, and people's bills. Caroline Lucas also argues that a 'win-win-win of climate policy is insulating homes' and Aurélien Lechevallier points to energy efficiency as 'the most important area in our current situation'.

Others have looked to invest the revenue in green technology. This is the favoured option of Dan Nocera, the inventor and scientist behind the Artificial Leaf. He argues that the real solution for climate change 'has very little to do with science now – I'm doing my job – instead it has to be political'.

Nocera's view is that carbon pricing could be used to 'jump-start' new, carbon-neutral and energy-efficient technological solutions, enabling them to compete with the fossil fuel industries that have already seen trillions of dollars of investment.

> If you solve the clean air problem, you solve the climate problem.
> Dawn Nakagawa, Berggruen Institute

Revenue could also be used to fund carbon-capture technology: if we priced carbon at the same rate it cost to remove, we would immediately solve the problem. In China, revenue from a carbon price, and air pollution penalties, have been invested in developing green technology.

Lisa Jackson says China is 'one of the great examples of how large countries – and economies – can shift to protecting the planet. The government recognized that its people want clean air to breathe, clean water to drink, a future for their kids.'

Though we need to be sensitive to its implementation, a price on carbon is the closest thing we have to an economic silver bullet for fixing climate change. In 2015, ahead of the Paris Climate Agreement, a letter signed by multiple Nobel Laureates, economists and former US cabinet secretaries urged negotiators to focus on widespread national carbon pricing. It claimed, 'This single policy change offers greater potential to combat global warming than any other policy, with minimal regulatory and enforcement costs.'

The complexities of protest

> An economist would probably argue that the
> most efficient way to reduce greenhouse gases
> is to put a price on carbon. But efficient is not
> always what can be attained from a political
> perspective.
> Christiana Figueres, *New York Times*,
> 'The problem with putting a price on the
> end of the world'

In application carbon pricing has been politically volatile and fraught with public backlashes. French President Macron's proposed increased tax on fuel in 2018 triggered the 'Yellow Vest' riots in France which saw nearly 300,000 people take to the streets, sometimes inciting violence and breaking property, and resulting in thousands of injuries and eleven deaths.

The Yellow Vest protests happened alongside a school climate strike in Paris. In the midst of these rising protests, I found myself at a climate-change-themed dinner, sitting next to President Macron's adviser, Aurélien Lechevallier.

'It's very confusing,' he admitted, commenting on the contrasting messages being sent by protestors: for more climate action, on the one hand, and against a fuel tax designed with the environment in mind, on the other.

Yet whilst the Yellow Vest protests have been used by some politicians to justify the failure of putting a price on pollution, Lechevallier says the issue was how the tax was implemented, penalizing ordinary people, rather than the ideas behind it.

Polls suggest that the Yellow Vest protestors were 'not against the efforts and the policies of the climate, it's just that for them, the day-to-day priority is the cost of living and struggle for life, and they want us to find a way to combine their way of life and the environmental approach'.

In 2012 Australia launched a national carbon price, offset by a reduction in income taxes. The effects were impressive: it was reported that coal usage dropped by 5 to 14 per cent, whilst renewable energy generation increased by 28 per cent. A 2014 report by the Australian National University estimated that the scheme had made the biggest annual reduction in greenhouse gas emissions in twenty-four years of records. Yet the media were very hostile to the carbon price, and there was a lack of clarity in how the revenue was distributed. The very same day that report was published, Australia became the first country to repeal the carbon price, and emissions immediately shot up. The vertical line in the graph (overleaf) represents the date of the repeal.

Meanwhile a carbon price proposal in New Zealand was abandoned, and in 2017 an attempt to introduce one in Washington was rejected. Are we, as a society, unwilling to pay for our pollution?

As Lechevallier knows painfully well, these examples point to failures of implementation, rather than failures of the idea itself. After stepping back from the proposed 'fuel tax', France did not abandon the idea:

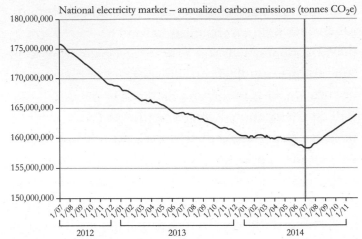

National electricity market – annualized carbon emissions (tonnes CO_2e)

The vertical line represents the date of the repeal

instead they re-designed it, shifting the focus from individuals to companies.

Wouldn't this trickle down to ordinary people, by making products more expensive? I asked. Lechevallier argued that if the system is transparent, consumers can see which companies are making environmental progress and reward them. This in effect should drive their product prices down, creating what he calls a 'win-win system for the environment but also for the prices and the fluidity of the market'. The key, as always, is transparency and communication.

There are many countries in Europe where carbon prices are effective. Finland was the first country to introduce one in 1990, with Denmark, Sweden and Norway quickly following. The UK government introduced a 'top-up' carbon pricing scheme, over and above the EU levels, which has been largely accredited with the drive away from coal, and emission reductions to levels barely seen since the nineteenth century. Carbon prices have been introduced or planned in over fifty countries, representing $40 billion in annual revenue and covering 20 per cent of annual global greenhouse emissions.

The International Monetary Fund (IMF) report on fossil fuel underpricing concluded: 'Efficient fossil fuel pricing in 2015 would have lowered global carbon emissions by 28 per cent and fossil fuel air pollution deaths by 46 per cent, and increased government revenue by 3.8 per cent of GDP.' Meanwhile, the OECD have called on countries to rethink energy pricing, direct subsidies and fossil fuel investment.

Josh Hoffman of Zymergen – who you may recall was making new materials like plastic *without* using oil – tells me he 'worries about the nature of transition' away from fossil fuels. He believes that 'an outright ban on fossil fuels would be a disaster today' and would cause a 'political mess and unforeseen consequences' in many parts of the world.

He believes instead we should 'create mechanisms that apply evolutionary pressure. A carbon tax is going to unleash a ton of creativity in people who think differently and have solutions that you and I can never imagine.'

A price on pollution, and reappraisal of subsidies, are not the only solutions, but they are low-hanging fruit for shifting the economy. They would render the true costs of energy more visible and give renewable energy a fighting chance.

The UK government claims to have reduced its emissions by 42 per cent whilst growing the economy by 72 per cent – with emissions peaking decades ago. However, when you include the impact of the UK's imported products, their overall emissions didn't peak until 2007. Still they have peaked – and that's worth celebrating.

Organisation for Economic Co-operation and Development: an intergovernmental economic organization with thirty-six mostly high-income, democratic member countries.

'The next three to five years are perhaps the most important of my lifetime,' reflects Caroline Lucas, who has dedicated her life to environmental politics. 'We must make bold and radical changes now. The key test for all of us is to focus on what is scientifically necessary, not just what is currently seen as politically possible.'

Guy Standing: 'Things like air, water and habitable earth. The essential aspect of the commons is that it doesn't belong to anyone in particular, it doesn't belong to the Crown, to the government, it doesn't belong to you or me: it belongs to all of us and none of us.' In 1217, the Charter of the Forest gave English people access and rights to the forests: defining the rights of the commons, and the rights of commoners. It had to be read out in churches four times a year for many centuries thereafter.

There is often a tension between a belief in big business or big government; the 'right' and the 'left'. Whether this transition will be led by significant investment from governments, or creating the policies, subsidies, taxes and incentives for business to lead the pivot – or a bit of both – often turns to our political ideologies. Yet these can become religions if we are not careful, and we could find ourselves so caught up in fighting about the best way to travel that we may miss the destination.

Meanwhile, let's not forget that third space: the often overlooked sector beyond business government and private interests – our commons. Can we defend and protect a *natural commons*?

Afraid of Kafkaesque government, or Wild West business, I try to explore the potential of both – not positioning myself as 'centrist' but rather believing that politics is more nuanced than a division between two camps, that good ideas cross-pollinate the political spectrum, and that we need both wings to fly.

Can citizens and their governments encourage capitalism to evolve to meet twenty-first-century challenges? Is this already under way?

> There is one and only one social responsibility
> of business – to use its resources and engage in
> activities designed to increase its profits so long
> as it stays within the rules of the game.
> Milton Friedman, *Capitalism and Freedom*

Milton Friedman's philosophy has been the religion, and legal underpinning, of capitalism for decades. As inequality escalates, and climate crisis calls, many are asking: can we rewrite the rules?

Even arch-capitalists are calling for a change in capitalism's direction. The former Starbucks CEO, Howard Schultz, has called for higher taxes on the wealthy saying the US faces a 'crisis of capitalism'. Ray Dalio, a billionaire hedge-fund manager, believes income inequality poses an 'existential threat' which could lead to a 'revolution of one sort or another' and advocates reforming capitalism: 'I believe that all good things taken to an extreme can be self-destructive and that everything must evolve or die. This is now true for capitalism.'

In 2019, the leaders of 181 of the biggest companies in the world appeared to break Milton's spell.

The companies represented – ranging from Amazon to Apple to JP Morgan – changed the official definition of the purpose of a corporation from profit

In 2020, the World Economic Forum released a new manifesto advocating stakeholder capitalism 'for a better kind of capitalism'. As part of this movement, 140 of the world's largest companies will begin reporting non-financial Environmental and Social Governance metrics to their stakeholders.

The finance industry is already changing course: in 2020, an internal letter from JP Morgan Chase economists warned of a 'global market failure' and advocated global carbon pricing. Having been the world's largest financier of fossil fuels, it joined Goldman Sachs in ruling out future investments in coal mining or arctic oil exploration.

maximization to pledge to 'lead their companies for the benefit of all <u>stakeholders</u> – customers, employees, suppliers, communities and shareholders'.

It was encouraging news. What can politics do to ensure these are deeds not words? We might soon see an 'accountable capitalism' Act come to pass, an international carbon price, a market-based Green New Deal, or the fundamental legal changes to the corporate game – away from 'shareholder primacy'– long championed by the social business movement.

The very real costs of climate breakdown might force these changes to happen soon. In April 2019, whilst Extinction Rebellion paraded and blockaded the streets of London, Bank of England governor Mark Carney warned that <u>financial institutions and investors</u> need to build the increasing costs and risks associated with climate breakdown into their modelling to 'help avoid a climate-driven sudden collapse in asset prices'.

What is clear is that we must encourage our representatives to make ambitious net zero carbon commitments, with immediate and coherent strategies, whilst retaining open minds to data, tactics and new discoveries along the way. This flexibility was President Roosevelt's original approach to the New Deal: 'It is common sense to take a method and try it. If it fails, admit it frankly and try another. But above all, try something.'

Chapter Ten
The Art of Listening

Direct democracy, solutions news and dreaming big

A rainbow of 193 flags flapped in the wind. Barbara Hepworth had carved stone circles in an otherwise phallic, mirrored landscape. A Japanese tree shuddered quietly. An African fertility bust stood trapped in a glass box. A black-and-white horse screamed, and a baby died, in Picasso's *Guernica*, near sleeping delegates.

Museum or dream?

I was visiting the United Nations in New York. One of the few positive outcomes of the Second World War, the UN is extraordinary. It has issued the Universal Declaration of Human Rights, fed over 90 million people, and saved millions of lives. From an environmental perspective, it is a key instrument for global strategy: generating the groundbreaking IPCC scientific reports, for example.

Yet the UN hasn't always been effective at creating peace and has been haunted by criticism and scandal. 'They talk too much,' said a lady I met on my visit, who had been working there for twelve years. 'Usually Nothing,' remarked a disillusioned refugee I met in a camp in Samos, Greece, wallpapered by blue-and-white UN printed tarpaulin.

Like all ideas, the UN is subject to the laws of entropy. It must continually adapt and evolve, especially as the pace of innovation and cultural exchange moves so quickly. This is true of all institutions and systems – including democracy itself.

'Democracy is the worst form of government except for all those other forms that have been tried,' Winston Churchill famously quipped, quoting someone else. There has been a lot of talk, over the last few years, of a crisis in politics. What's certain is that even in a representative democracy, politics can feel distant, maddening, irrelevant or self-obsessed.

Why, might we ask, is street protest necessary? Why do people like XR and the School Climate activists have to chain themselves to trees, glue themselves to Parliament's glass walls, or sit cold in jail cells, in desperate attempts to make their voices heard? Why are they not already better represented?

The art of listening

What holds back our political system from action on climate change is the lack of true democracy embedded within it.
Caroline Lucas, Green Party MP for
Brighton Pavilion

Which of these two or three people are most like you? We get to answer this ridiculous question only once every four to five years, handing our country's destiny to the winner. It's no wonder people get so infuriated by politics.

In the twenty-first century, can we not do democracy – the art of listening – better?

When XR asked for a Citizens' Assembly on climate change in 2018, many had not heard of the democratic construct that seeks to bring the voices of ordinary people into the political discussion.

It had already proven a useful ally to the climate in Ireland. In 2016, Ireland instituted a citizens' assembly wherein ninety-nine citizens were randomly selected as a broad representation of the Irish electorate, to debate important policy issues, informed with facts and impartial expert advice. Irish environmental writer John Gibbons said the citizens' assembly led to 'the strongest cross-party political statement of intent on climate action'.

After announcing the 2050 goal of net zero, the British government seemingly ceded to XR's demand, and pledged to convene citizens' assemblies on climate change to help define popular ways to meet the goal.

Old dreams

The progress from an absolute to a limited
monarchy, from a limited monarchy to a
democracy, is a progress toward a true respect for
the individual. . . . Is a democracy, such as we know
it, the last improvement possible in government?
Henry David Thoreau,
Civil Disobedience, 1849

Thoreau believed that democracy could evolve towards increasing people power, yet over 170 years later, our main political methodology has barely changed. We have made politics more inclusive, allowing women and people of colour to vote, but the fundamental voting dynamics of government remain the same, even as we have watched digital disrupt most other sectors.

If we can bank online, why can't we vote online? Why is the political establishment so reluctant to embrace digital? Why are election days not national holidays and why can't sixteen- and seventeen-year-olds vote in such as like the US, UK and Canada? Why is voter registration not always automatic? Who is the suppressed political participation (especially of the digitally savvy youth) really serving?

Truth, like a bubble, rises. With digital, this landscape is changing. Whilst we have explored the ways in which digital creates new and unprecedented threats to democracy, it also offers to positively transform our engagement with politics, opening new pathways to increase awareness, dialogue and transparency.

Estonia has offered digital electoral voting since 2005. Websites are already being used to gather citizens' views, and make representatives' actions more transparent. Many national governments, and the EU, have also built their own digital petitioning portals to gauge public consensus: promising to review and respond to large petitions.

Digital is also reviving a very old dream: direct democracy.

The idea of enabling ordinary people to vote directly on individual laws has been around since the Athenian democracy of the fifth century BC (albeit one exclusive to certain men), and was part of Gandhi's proposal for Swaraj: localized, participatory democracy that gave equal power to all citizens.

Truly representative, direct versions of democracy enable and empower citizens to have a closer relationship to governance, and focus on the nuance of the issues at stake. They also help to move us away from the dangerously divisive false narrative of bi-partisan politics.

Across Europe, 'Pirate Parties' have been formed with the headline promise of championing direct, digital democracy, alongside reforming copyright and patent laws, opposing state surveillance and evangelizing open data. Pirate Party members have won increasing political power: being elected as European Members of Parliament (Sweden and Germany), Members of Parliament (Germany and Iceland), and as a Pirate Party mayor (in the Czech Republic).

Meanwhile, in Switzerland, direct democracy has been the mainstream political model for over a century. Since 1891, it has offered its citizens a hybrid of representational and direct democracy. All of the 8.5 million citizens who are over eighteen years old (including women only – shockingly – since 1971, and 1990 in one canton!) can propose new policies, and call for referendums to veto planned laws, seeing four popular referendums on a selection of issues each year.

The Swiss have made some surprising choices, sometimes seemingly against their self-interest: 66 per cent voted against having two weeks' extra holiday pay, and 58 per cent voted to move away from nuclear power towards renewables even though this would have a higher impact on their energy bills.

A report in 2016 found that the Swiss are nearly twice as likely to trust their government compared to the average across OECD countries. My friend Osiris Roost in Switzerland says, 'People don't complain about politicians here, because they don't have a lot of power.'

New visions

Too often, observers deem Switzerland an oddity among political systems. It is more appropriate to regard it as a pioneer.

Kris Kobach, *The Referendum: Direct Democracy in Switzerland*

'They have to take to the streets to be heard!' was Argentinian activist and technologist Pia Mancini's rallying call, when she founded a digital direct democracy platform to try to upgrade democracy to the twenty-first century.

In 2012, she and Santiago Siri launched 'Democracy OS' – an app and open-source technology that sought to encourage citizens' political engagement through discussion and online voting. They formed a political party – the Partido de la Red (Party of the Net) – which ran for office in Buenos Aires, on the pledge that elected officials would simply be a human conduit for enacting whatever the citizens voted for online: 'We had one rule,' Siri says. 'Obey the Internet.'

They wanted to disrupt the political system from the inside. The fledgling party captured a lot of public and media attention, especially when they trundled a large wooden Trojan horse through the streets of Buenos Aires and left it outside Congress, surrounded by placards calling for politicians of all parties to step aside and make way for the will of the people. The Internet Party received 22,000 votes in the elections for the city government, just over 1 per cent of the total vote and about five times what most new parties receive.

They planned to keep building on their initial momentum but according to Santiago, were intimidated and asked for bribes: 'Changing the system from within was not going to happen. The system was going to change you first.'

Santiago and Pia now run Democracy Earth, an incubator for digital democracies, experimenting with ideas such as liquid voting, quadratic voting and blockchain technology. 'Our political system can be transformed, not by subverting it, by destroying it, but by re-wiring it with the tools that the Internet affords us now,' says Pia.

Their work lands on fertile ground with many technologists looking to redesign democracy. Open-source proponent Tim O'Reilly suggests giving

Lies somewhere between representative and direct democracy: individuals can vote directly on issues, or choose to allocate their votes to other people to vote on their behalf (such as a specialist on the topic, or a trusted knowledgeable friend).

The idea of products that grant permission to be used, modified and shared. Although largely software-orientated as a term, it has been applied far more broadly. One can find open-source buildings, soft drinks, films, music, religion and furniture.

politicians 'algorithmic regulation' to guide their decision-making, so they can lead with a more entrepreneurial, flexible and data-driven spirit.

Meanwhile, MIT professor César Hidalgo proposes building personalized Artificial Intelligence robots that are super-informed on issues, and know our beliefs and values. These robots would then become our personal representatives to vote on legislation for us, allowing us to choose how involved we want to be.

Santiago Siri believes that the digital evolution of democracy, giving people more sovereignty, is inevitable. 'Everyone has a phone in their pockets. My generation and the one that comes after me are shaped by the Internet. What kind of politics will we see emerging from the Internet?' He thinks we will see shifts in the next few decades, although he admits, 'The closer to the problem you get, the more confused you become.' Upgrading democracy, like political protest, is a long – but arguably necessary – game.

Cleaning the media mirror

A healthy democracy depends on having an informed electorate, which means holding the media, social media and politicians accountable to the truth. This becomes increasingly critical if moves are made towards direct democracy.

When you talk about regular referendums in the UK, people often get uncomfortable. 'The last thing we need is more referendums,' said a friend, baulking at the idea post-Brexit. The 2016 referendum on Brexit was full of misinformation; it was divisive and hostile. Yet was it the uniqueness of that referendum – the fact that it was a privilege not regularly given – that caused much of the drama and distortion?

In Switzerland, Osiris says, there are also attempts at misinformation around referendums, but the government makes efforts to issue 'completely neutral information' and 'the fact that everyone is interested and well informed, means misinformation has mostly no effect in the outcome. I am always talking to a lot of people, young and old, asking them opinions about voting questions. It is always a topic at lunch tables.'

As we saw in Chapter Four, a series of challenges confronts our contemporary media landscape: advertising and data manipulation, fake news, mass surveillance, bots, media monopolies, and filter bubbles that drive polarization, underpinned by predominantly for-profit business models, and private interests.

The media has become an anamorphic, funfair mirror to our collective consciousness which in turn distorts our ability to have a healthy democracy.

Yet there are people trying to clean it.

Politicians are talking about regulating – possibly even breaking up – social media monopolies. In the fight against fake news in the run-up to the 2017 French elections, a coalition of media companies including Google and Facebook launched CrossCheck, to help identify and fact-check content, and label the dubious material. Facebook and YouTube are linking to Wikipedia to be the 'good cop' of the Internet, to help give context to potential fake news.

In spite of these serious challenges, we ought not to lose sight of the fact that the digital landscape has also opened up unprecedented access to information. Mobile phones, by putting a camera in everyone's pocket, have empowered citizen-led journalism, allowing the first-hand documentation of events in ways we have never seen before: a <u>panopticon</u> society that can work for people, as much as it does for states, as seen in the civilian documentation of police abuse, and the Black Lives Matter movement.

A Reuters report found that social media users in the US were more than twice as likely to see news from different political perspectives as people who didn't use social media. If you do some research and dig around, you can find surveys, facts and disparate opinions, or imperfect but helpful Wikipedia articles on any global issue you care to know about.

A theoretical circular prison proposed by eighteenth-century philosopher Jeremy Bentham whereby all prisoners can be seen by a single guard in a central tower without the inmates knowing whether or not they are being observed. The notion of possibly being watched ('a sort of invisible omnipresence' as Bentham put it) shapes the prisoners' behaviour. Parallels have been drawn, understandably, with the growth of surveillance culture.

An honest media

If you don't read the newspaper, you're uninformed. If you read the newspaper, you're misinformed.

Mark Twain

The media not only reflects our cultural zeitgeist, it also shapes it. It is essential we are aware of the costs, risks and threats we face as a society: to understand the suffering of others, bear witness to tragedies and investigate corruption. We need to know about those alarming scientific reports on climate and biodiversity, as per X R's central ask of the media to 'tell the truth'.

Yet it is equally important that we are made aware of all the solutions being explored, and the steps that communities and politicians are taking – and can take – towards making positive change.

Solutions-based journalism is 'rigorous journalism that reports critically on tangible progress being made, in order for us to understand how issues are being dealt with,' says Jodie Jackson, a research partner for the Constructive Journalism Project and author of *You Are What You Read*. News sites such as the *Guardian* and the BBC have created 'positive news' sections and there are a rising number of news sites dedicated to positive journalism.

I like The Happy Broadcast and Positive News,

'I would like to see the media industry report on strength as it does on weakness, on successes as it does on failures, on human excellence as it does on human corruption and scandal, on solutions as it does on problems, and on progress as it does on recession,' she says.

Solution-based news would help turn the tide against the avalanche of fear-based information we receive. This is crucial not only in the service of standards of objective truth-based journalism, but also because consuming only bad news is like eating junk food.

Research shows that 'prolonged exposure to bad news over long periods can have detrimental effects on moods, attitudes, perceptions and emotional health'. Negative news leads to fear, division, distrust, anxiety and a de-sensitization to the world.

Solutions-based news, or constructive journalism, arguably also offers us a more accurate picture of reality: global poverty, maternal and child mortality have dropped; social spending and life expectancy have steadily risen; literacy and liberal values are up; our democracies may not be perfect but there are more of them; we are curing diseases, and making daily breakthroughs in science and technology.

> If you had to choose a moment in history to be
> born, and you did not know ahead of time who
> you would be – you didn't know whether you
> were going to be born into a wealthy family or
> a poor family, what country you'd be born in,
> whether you were going to be a man or a woman
> – if you had to choose blindly what moment
> you'd want to be born, you'd choose now.
> Barack Obama, in a speech given at
> Goalkeepers conference, New York, 2017

Psychologist, linguist and Harvard academic Steven Pinker uses data to paint a picture of our improved collective state of affairs, in his book *Enlightenment Now*. He argues that people often do not recognize how much progress we have made because the media increasingly emphasizes negative news under the general mandate, 'If it bleeds, it leads.'

'The unsolved problems facing the world today are gargantuan, including the risks of climate change and nuclear war,' says Pinker, 'but we need to see them as problems to be solved, not apocalypses in waiting, and aggressively pursue solutions.'

He has ruffled many feathers with his techno-capitalist optimism, and I must say I struggle with some of it. His claim that this is likely the most peaceful time in human history offers cold comfort to the people living in war zones today, and doesn't take account of the violence being inflicted on other species or the natural world itself.

Yet I do think there is a lot we can learn from absorbing the data, and questioning how we perceive and understand this moment in which we live. In many, rarely celebrated ways, things are truly great.

We have further to go, but a media that puts our challenges in the context of our successes and possibilities, will give us the courage to make the collective, cognitive shifts needed. Whilst we wait for positive news sites to grow, we can take the initiative to research and look for solutions. Knowing that they exist, and are possible, will help us imagine better futures, and then vote, act, buy or live them into reality.

The power to imagine

Everything you can imagine is real.
Pablo Picasso

The same week that I visited Picasso's *Guernica* in the UN building in New York, I encountered another

UN. This second one was much smaller with an asymmetrical curvaceous body. Silver, with rippling Yves Klein blue, yellow and purple stars.

Congolese artist Bodys Isek Kingelez's fantastical, celebratory cardboard sculpture *U.N.* was constructed in 1995, on the fiftieth anniversary of the real one. It makes an oblique and optimistic reference to the UN's peacekeeping mission during civil unrest from 1994 in the Congo.

It was simultaneously bright, gaudy and fun; and modest and fragile. According to its maker, 'In this palace, peace is an indispensable tool for the democracy of nations.'

Originally a school teacher in Zaire (now the Democratic Republic of Congo), Kingelez must have seemed like little more than a local madman when he started making his paper sculptures of utopian cities using a Gillette razor, paper and glue. His utopian visions – an alternative to the World Trade Center towers, a hospital for the AIDs crisis, an antidote for Hiroshima, a 'peaceful city where everybody is free' and 'a city that breathes nothing but joy' – used humble materials, tea packages and soft drinks cans to make elaborate designs.

Kingelez reworked capitalist debris – literally and symbolically – into positive, playful and almost child-like spaces. He joins a long line of visionary artists and architects who have imagined new visions of society.

What is now proved was once only imagined.
William Blake, 'Proverbs of Hell'

284

Imagination is what differentiates humans from other animal species, and it offers us the first step in change.

We might find that many of our institutions – the UN, the EU, capitalism and democracy itself – are imperfectly formed, yet they were all born out of our ancestors' imaginations, and are asking always to be reimagined.

How far can we stretch our imaginations to evolve them?

Perhaps it is even worth searching yourself for the visions of society you would like to see come true: for the better world you know in your heart is possible.

'Without a model, you are nowhere. A nation that can't make models is a nation that doesn't understand things, a nation that doesn't live,' said Kingelez. 'If you succeed in building a model, you visualize what is living inside you, so that the outside world can adapt it, study it, discover it, see it.'

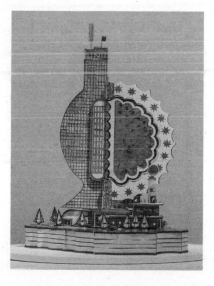

Bodys Isek Kingelez, *U.N.*, 1995.

PART FOUR

The Roots

Are the real solutions inside us?

Insanity is doing the same thing over and over
again, and expecting different results.
                                    Albert Einstein

So far in this book I have looked at the routes being
pursued to meet our environmental and social challenges, in the hope that we can evolve capitalism and
its connected facets of politics and technology.

But can you ever change the system from inside?

Or is everything I have said so far deeply naive?

More than routes to travel, do we need to examine
our roots?

Might we need to dig deeper, and rethink how we
actually think?

Over the years, I have come increasingly to sense
that the most important part of a move towards sustainability begins in our hearts and minds: in the way
we see the world, one another, and our place within
the whole.

In our collective beliefs, mythologies and dreams.

Throughout history there have been numerous
alternatives to the system we live in, and wildly imagined visions of our present and future.

There was a time when most humans believed that
the Earth was flat, and that an old white man in the
sky ruled supreme. Now we are more likely to believe that we can own patches of land and even other

people's hearts; and that money, contracts and bank accounts are real.

> Ever since the Cognitive Revolution, Sapiens have
> thus been living in a dual reality. On the one hand,
> the objective reality of rivers, trees and lions; and
> on the other hand, the imagined reality of gods,
> nations and corporations. As time went by, the
> imagined reality became ever more powerful, so
> that today the very survival of rivers, trees and
> lions depends on the grace of imagined entities
> such as the United States and Google.
> Yuval Noah Harari, *Sapiens*

When we realize that how we think about the world doesn't represent truth or reality so much as the latest version of a story we've all bought into, then we can start thinking about the next chapter we want to write.

Who are the radicals and outliers with a different take? What can we learn from the oldest, most sustainable communities on the Earth? Can we reconcile those learnings and mindsets with modern inventions? What simple things are we already doing ourselves that might point us in a better direction?

My journey began with diamonds, but I realized along the way that perhaps the solutions we need are less about ethically sourcing diamonds, or even making diamonds from carbon dioxide emissions, and more about questioning the desire for them at all.

Chapter Eleven
Who Cares Wins?

Community, the gift economy and choosing kindness

Re-examine all you have been told . . . dismiss
whatever insults your own soul.
Walt Whitman, *Leaves of Grass*

One week in 2010, two emails arrived in my inbox,
offering what seemed to be a playful test from the
universe – but transpired to be a definitive fork in
my path.

I was presented with an odd choice: did I want to
walk in a high-profile fashion show, or visit a Burmese
refugee camp?

As Halloween dawned, I was on a flight to Thailand to visit the Thai–Burmese border, as part of the
fundraising trip to draw attention to a refugee camp
that is so old it had almost been forgotten. I'd taken
a week off university, and was jammed into an aisle
with the charity representative and my friend Kate
Tomlinson, having a conversation about the 2008 eco
nomic crisis.

We were marvelling over the fact that when de-
prived of money, a society completely falls apart:
all the *real* resources – skills, objects, time – are still
there, yet people become largely helpless. It seems our
society has become slaves to a tool we invented.

'Could technology,' we conjectured, 'connect people to trade directly, without money in between?' The idea seemed to bubble up through our conversation, like a gift, until it landed on the fairly anti-climactic suggestion of 'a website' – and then we sat in silence for a moment, pondering. *Why does that not already exist?*

Our tentative idea felt all the more poignant over the next week as we spent time in the camp: one of nine that run along the Thai–Burmese border, makeshift homes to nearly 100,000 people. They were set up in 1984, but the Thai government has resisted permanence in the camps, so basic infrastructure is missing. The delicate bamboo-frame homes have been built and rebuilt, over and over again, for more than thirty years. People have passed from cradle to grave, never exploring the world beyond their mile-wide camp.

Most of the refugees are ethnic minorities who fled persecution in Burma. Food is rationed and simple, and choices are limited. Yet in spite – or perhaps because – of the difficulties that these people face, there is extraordinary collective spirit and resilience. With hardly any money, this community functions through a network of small favours and interdependence. Women exchanged stories at the ration lines, and kids played by the stream. Many of the people I met seemed connected and happy in ways I don't see so often in my home town. Though I had gone there to 'help', I was moved to see how much I had to learn.

Paradise

The 2008 financial crash philosophically and politically stirred our society. Though people have been questioning 'the system' for many years, a whole generation of young thought-leaders – many of whom I have discussed in this book – make claim to 2008 as a radical turning point in their thinking.

The crisis revealed the insecurities in our system: cracks begging to be filled with new visions. Although governments bailed out the banks, the Occupy tents were dismantled and business as usual was resumed, bolstered by harsh austerity, yet the crisis had planted seeds in the minds of many, now yielding wild flowers worldwide.

Crisis often provokes growth of a different kind. Necessity is the mother of reinvention. After the economic collapse in Argentina in 2001, widespread barter networks developed, as a mechanism for survival. It is estimated that between 4 and 7 million people participated.

In a similar vein, in Spain in 2012, when unemployment averaged 25 per cent, there were 325 TimeBanks allowing people to exchange their time – and alternative local currencies appeared in different towns. In Greece, when unemployment topped 20 per cent in 2011, the town of Volos turned a local vet's office into an informal community centre, and launched an online barter network using a new currency.

Time is a currency, and an hour of my time is equal in value to an hour of your time. To date, one TimeBank website, hourworld. org (its motto: Peace, Abundance and Sharing through Equality in Exchange!), has logged over 2 million hours of service.

The conditions under which these systems arose were deeply anxiety-inducing and they were more a means of survival than an experiment in alternative ways of living. Nevertheless they reflect the resourcefulness that people exhibited, and in different circumstances we might learn from them how to build resilience into our own systems.

Now, more than a decade after the 2008 financial crash, in the midst of an environmental emergency and ever haunted by the possibility of recession, is it high time to question some of our deepest assumptions?

What would our world look like if money and the need to earn it played less central roles in our lives? If economic growth and GDP weren't our countries' primary metrics? Can we foster a shadow economy of gifts and favours? What could we learn about trust, community and sustainability from groups who are exploring these questions? Would these shifts make us happier? Would they better insulate us to deal with crisis? And which elements of this vision are already all around us, waiting to emerge and grow?

The currency of trust

Trust involves the delicate juxtaposition of
people's loftiest hopes and aspirations with their
deepest worries and darkest fears.
Morton Deutsch, *The Resolution of Conflict*

I got home from the trip to Burma and continued with my degree, in my final year of reading History of Art. As I worked my way through the course, this idea – whether technology can play the role of money – shadowed me. It seemed so simple and so obvious, yet try as I might I couldn't find anything online offering it in a way that spoke to me, or operating at scale.

I could see that platforms in this space were bubbling up: from the emergence of Bitcoin, TimeBanks, collaborative consumption (now better known as the 'sharing economy'), to the early designs of a niche website called Airbnb. I could see that a movement was brewing.

A fellow student told me about something called the gift economy, and directed me towards its iconic thesis, *The Gift*, written in 1925 by anthropologist Marcel Mauss. I found it in my college library and read, enthralled.

Mauss had studied different tribes around the world that structured trade relationships on the concept of gift-giving, rather than through the transactional relationships and exchange mechanisms with which we are so familiar.

For many of us, the concept of private property – and therefore exchange – features so pervasively in our reality that we barely pause to question it, yet private property is of course an artificial construct that was, at one point, invented and enforced.

The first and most famous of all cryptocurrencies – the Internet-based medium of exchange that uses blockchain technology to allow for decentralization from government banking systems. Since the release of Bitcoin almost 2,000 other forms of cryptocurrency have been launched with all sorts of core ideas and guiding principles behind them including Auroracoin (created as an alternative currency for Iceland), Dentacoin (made by dentists for dentists), Mooncoin (with a total coin supply based upon the average distance from the Earth to the Moon) and Dogecoin (a cryptocurrency featuring the likeness of a Shiba Inu dog).

The first man who, having fenced in a piece of
land, said, 'This is mine,' and found people naive
enough to believe him, that man was the true
founder of civil society. From how many crimes,
wars, and murders, from how many horrors and
misfortunes might not any one have saved
mankind, by pulling up the stakes, or filling up
the ditch, and crying to his fellows: Beware of
listening to this impostor; you are undone if you
once forget that the fruits of the earth belong to
us all, and the earth itself to nobody.

Jean-Jacques Rousseau, *On the Origin of the
Inequality of Mankind*

A gift-economy practice in the Kula ring chain of islands in Papua New Guinea. Participants travel by canoe, sometimes hundreds of miles, to exchange shell necklaces (which are traded clockwise around the ring) and shell armbands (which are traded counter-clockwise). The gifts hold neither financial value nor any particular use but instead are traded to enhance social status and strengthen

According to most anthropologists, the gift economy, dependent on mutual reciprocity and trust as a currency, is the most ancient way that *Homo sapiens* interacted with one another. From gatherer-hunter economies, to the <u>Kula ring exchanges</u> in Papua New Guinea, this was the norm long before barter began, gold was mined from the earth, or silicon enabled numbers to appear on screens.

My imagination was gripped by the radically different alternatives to our society that Mauss described. One of his investigations was of potlatches organized by indigenous communities in the Pacific Northwest and US. 'Potlatch' means 'to give away'. A potlatch would typically mark an important event, such as a birth, wedding, death or coming-of-age ceremony, involving communal feasting and reciprocal gift-giving. The potlatch was the tribes' main economic activity, and social reputation was established according to who could give the most. There was also significance in receiving: chiefs would establish their power through gifts to visitors – by accepting the gifts, visitors would be showing their approval.

Mauss argued that these communities believed in an almost spiritual force that would bind the giver and the receiver. Gifts were always in motion, and an implicit trust in their being reciprocated kept the relationships among the community alive.

Gift economies required and offered a deep trust that one's needs would be met. In the Sepik Coast area of Papua New Guinea, families would form alliances with other families through the practice of giving objects such as baskets, wooden bowls or tobacco. When they travelled between towns, they would bring gifts to their contact family who would house and care for them. Families might have up to seventy-five contacts, cross-generationally, and through the vast quantity of gifts exchanged, their needs would generally be met.

My reading wound me back to the journey I had begun with Dr James Suzman all those years earlier in Botswana, trying to learn about, and from, the San. In twenty years of studying San communities and their histories and practices, James concludes that they offer the most sustainable model of human living, 'if the ultimate measure of sustainability is endurance over time'. Nine tenths of human history was lived as gatherer-hunters.

relationships between the large number of participating islands. When you think about it, a similar though more random logic can be found in the (often bizarre) gifts exchanged between world leaders. In 2013 Barack Obama was given twelve scented votive candles and a tea infuser in the shape of a penguin by the Sultan of Brunei.

James argues that the success of the San is rooted in 'having few needs that are easily met', a culture of sharing (i.e. the gift economy), a disdain of waste, a lack of hierarchy (practised through 'insulting the meat'), and a deep trust in providence – that the environment always provides. (This sentiment is echoed in the contemporary gift economy experiment, Burning Man, where people often declare that 'the playa provides'.)

According to Mauss and James, these communities seem to harbour a sense of abundance; a trust that everyone's needs will be met, and a measure of wealth in relationships. In London where I grew up, I was hungry for community, poor in connection in ways I didn't even realize. I also felt this constant sense of fear surrounding our security: perhaps because my mum was poor, perhaps because fear is the natural offspring of the competitive nature of our system, perhaps because we were missing community.

Reading Mauss, I found myself wondering, could a movement back towards community and sharing be replicated in our culture, in which we rely on vast anonymous networks of people to navigate our daily lives? Is it natural to pass tens, hundreds – sometimes thousands – of people each day, and rarely say hello? Why are mental health issues so high in modern 'affluent' societies? Are sharing and mutualism our natural ways of being?

I wanted to see if technology could re-catalyse deep-rooted patterns of human behaviour in cities, to help build bridges between individuals and encourage the social, community-rich behaviour you

Every year 70,000 people gather in the Nevada desert for nine days of alternative living guided by principles that include gifting of everything from sushi to massages (the only things available to buy are coffee and ice). Is Burning Man a privileged party, a transcendental experience, or a transformative way of life? I've been twice, and it's a mad mix of it all, but truly inspiring to see how happy and generous people are in a culture of giving. Regional shoots exist around the world such as AfrikaBurn.

find in pre-capitalist tribes, or smaller communities around the world.

I also wanted to see if we could grow the culture of sharing, which *does* already exist in our society, to coexist on a stronger footing with our monetary one – for reasons of social and psychological value, sustainability, and also as an insurance policy against the next economic crisis we might meet. As our climate crisis escalates, this seems more essential than ever.

Many people thought I was crazy and/or naive.

But it was worth trying.

The Spanish word for beach is used affectionately to describe the dried-up lake bed that Burning Man sits on. A playa chicken is the mythical belief in vicious, desert-dwelling chickens, that are responsible for any inexplicable occurrences.

Lessons from nature

We are the people of the wild boar – the people
who walk in bands – our strength is in our
collectiveness, our community. One wild boar
on its own is fearful and weak, when he is in his
band he feels very powerful.
Putanny Luiza, the Yawanawa tribe

Just before I left university, I looked up economics professors and was directed to the dark and dusty room of an expert in global economics. I enthusiastically explained the idea for an online gift economy. He stared at me, baffled, for a long time, then asked, 'But why would anyone do anything for anyone else for free?'

I tried again to explain how gift economies have, could and might work.

Meat is shared among the !Kung after each hunt but the hunter who provided the kill is invariably insulted by everyone else. Far from bad manners, the ritual insulting of all kills, whether a meagre haul or a veritable bounty, is practised to avoid the possible upsetting of an egalitarian society through big-headedness and notions of being better than anyone else.

'Ah!' he eventually said, his face breaking into a wide smile. 'Yes, I get it. It's like the way birds fly in a V-shape!'

He was referring to the way in which flocks of flying birds shelter one another from the wind, taking turns to be in the tricky spot at the front, where the wind resistance is higher. Like a termite nest, beehive, galaxy spiral, or the way penguins huddle together to stay warm, there is no governing structure – no one determining which bird goes next or which penguin gets to stand in the centre – yet a pattern emerges through a decentralized and spontaneous process. They naturally cooperate.

Whilst I was in California, interviewing people for this book, I went on a local quest to find an albino redwood. Like most souls, I've long been enchanted by the magnificence of redwoods, and the prospect of catching sight of a very rare ghostly one drove my search. There were nine to discover within the cathedral-like forest of ancient redwoods, some 2,000 years old, one so cavernous it once functioned as a hotel bedroom. A volunteer working at the park led us through the trees and along a railroad, towards a white, miniature redwood, grasping the edge of its giant sister.

Finding the albinos was visually fairly disappointing, to be frank – like a small white fern – but discovering their story wasn't. The generosity shown from normal redwoods to albino redwoods has perplexed scientists for at least a century. Without the ability to photosynthesize, the albinos require normal redwoods to survive, drinking sugars through their root networks, earning the nickname 'vampire trees'.

Trees have been found to communicate underground while redwoods specifically share nutrients amongst themselves via their root network throughout winter and spring. Come summer, they stop doing this with all except the albino. Why?

Baffled by this question, local biologist Zane Moore started researching the role of albino redwoods in the forests, and found that they contain twice the levels of toxic heavy metals (cadmium, copper, nickel) compared to normal redwoods: 'It seems like the albino trees are just sucking these heavy metals up out of the soil,' Moore said. 'They're basically poisoning themselves.'

According to his research, the albino redwoods are not vampires but angels; not parasites but filters; friendly ghosts sacrificing themselves to maintain soil health for the forest.

Often described as an underground Internet or the 'wood-wide web', trees and many other forest plants have been found to use a vast network of mycorrhizal fungi that live within and around their root systems and allow them to do such miraculous things as share nutrients and even warn of incoming pests or predators.

Kindness prevails

Looking to future generations . . . virtue will be triumphant.
Charles Darwin, *The Descent of Man*

The albino redwood is not unique. Nature is full of examples of symbiosis. If you go back to the very first singular organism on planet Earth, the beginning of all evolution, you find an example of both competition and cooperation.

Our dependence on the natural world for oxygen, nutrients and energy goes without saying. Inside our own bodies we host trillions of microorganisms, and at least 1,000 different species of bacteria, which we couldn't survive without.

Whilst Charles Darwin's name typically comes up in discussions of man as a competitive, self-interested creature, through his famous theory of 'Survival of the Fittest', Darwin also wrote extensively about the vital role of cooperation in evolution. In *The Descent of Man* he described loving and sympathetic behaviour in multiple species including elephants, dogs, baboons and pelicans as well as humans, to argue that natural selection favours cooperation. In this reading, giving is not self-sacrificial: we cooperate because it helps us. The flying birds and the penguins take turns to enable the whole group to survive.

Perhaps because of evolution, it turns out that humans are neurologically hard-wired to give. Giving releases hormones such as <u>oxytocin</u> which produce a natural high. In one study people were given the choice to keep $128 or donate it to charity; the researchers found that those who donated it activated the regions of the brain associated with positive reward, pleasure, trust and social connection. Giving is also good for us: volunteering has been shown to reduce levels of stress-related hormones, anxiety and depression.

Further studies have suggested that humans will be generous even to anonymous strangers when there

Due to its role in behaviours such as sexual arousal, trust and mother–infant bonding, oxytocin sometimes goes by the nicknames 'love hormone' or 'cuddle chemical'. Its less catchy chemical formula is $C_{43}H_{66}N_{12}O_{12}S_2$.

is no guarantee of reciprocation: seemingly irrational behaviour according to traditional economists. A group of psychologists consequently concluded that generosity is a built-in survival strategy: 'Human generosity, far from being a thin veneer of cultural conditioning atop a Machiavellian core, may turn out to be a bedrock feature of human nature.'

The online project Wearelucky, which gave strangers $1,000 and asked them to do something 'good' with it, found similar effects. One participant, who used the money to help a friend adopt a child, pay his mother-in-law's medical bills and tip a waitress $100 after taking his kids to lunch, said, 'I did think about spending the money on myself but I felt doing good is about giving. It has been so much more fulfilling to give it away than spend it on myself.'

> Every man is a golden link in the
> chain of my good.
> Florence Scovel Shinn,
> *Your Word is Your Wand*

Generosity is also contagious. A Harvard study demonstrated that when one person behaves generously, it inspires others to behave generously towards different people, and that altruism could thus spread through three degrees of separation. Likewise, Marcel Mauss compared the sense of gratitude to a <u>debt</u>: inspiring the recipient to pay it back or forward. Have you ever received a gift from someone, felt grateful, and then felt like you need to respond in kind somehow?

Gifts weren't all rosy in Mauss's view. He looked at the shadow side of giving and how – taken to its extreme – it could potentially create power imbalances. The Inuit are often quoted as saying 'gifts make slaves like whips make dogs'.

Muni ~1575–85;
back forma-
tion from Latin
mūnificentia gener-
osity, munificence,
equivalent to
mūnific(us) generous
(*muni-*, combining
form of *mūnus* gift
+ *-ficus* -fic) + *-entia*
-ence.

The reciprocal nature of the gift is what helps create community. In fact, the word '<u>muni</u>' – central to 'com*muni*ty' – derives from the Latin word for 'gift'. Community has been found to play a central role in human health and happiness. A 2006 University of California study showed that out of 3,000 women with breast cancer, those with a large network of friends were four times more likely to survive than those with fewer connections.

In Sicily, there is a town which is known as a 'blue zone' because it has one of the highest concentrations in the world of people who live to a hundred years old. Researchers have found that the primary and secondary reasons for their extra lifespans are social integration and relationships. Psychologist and author Susan Pinker argues that three is the magic number: the number of close relationships needed to ensure good health. To put it in practical terms: make real friends (at least three) and keep them for as long as you can. Create your village and take care of it.

Just as we all require food, water and sleep to
survive, we all need genuine human contact.
Susan Pinker, *Guardian*, 'Why face-to-face
contact matters in our digital age'

Utopia made real: everyday sharing

As I looked back at our Earth from the orbital
perspective – this stunning, fragile oasis that has

been given to us, and has protected all life from
the harshness of space – I saw a world where
natural and human-defined boundaries shrank.
I saw a world becoming more and more interconnected and collaborative – a world where
the exponential increase in technology is making
the 'impossible possible on a daily basis'.
Ron Garan, *The Orbital Perspective*

If organizing our society around the principles of
trust and generosity feels like a utopian dream, lovely
but out of reach, it might be inspiring to recall that the
welfare state and the NHS, along with equivalents
in countries all over the world, must once have felt
similarly unlikely to those people who were advocating them.

It might also be worth reflecting on the fact that,
according to the British government, doing things for
each other for free is already as big as GDP in the
UK. We all participate in this shadow economy to a
greater or lesser degree: think of the household work,
caring, child rearing, favours amongst friends, families, colleagues or neighbours; Thanksgiving lunches,
helping a friend move house, Christmas gifts and giving a stranger directions.

Meanwhile, sharing already manifests in myriad
ways that many of us take for granted: libraries, buses, roads and parks, power systems and
water supplies, charity donations and shops, bul-
letin boards, community gardens and universities.
Think also of the generous spirit that comes alive

in response to natural disasters, and also at times of communal festivities – such as <u>Mardi Gras</u> and carnivals.

Then there's classical philanthropy. Hundreds of billions of dollars are donated every year, and trillions of dollars run through non-profit organizations. Many people volunteer their time or work on the front line of crisis. In Samos, Greece, in 2016, I witnessed first-hand the international grass-roots response to the refugee crisis. Tens of thousands of people gave their time to help, and millions of pounds' worth of items – clothes, shoes, food – were donated.

Meanwhile, if you look at your day-to-day reality – and I don't mean what you read in the news – the majority of people are trustworthy, and there are far more acts of kindness than of cruelty. As banal as it seems, kindness is the status quo.

> Not I, not I, but the wind that blows through
> me! . . . If only I am sensitive, subtle, oh, delicate,
> a winged gift!
> D. H. Lawrence, 'Song of a Man Who Has
> Come Through'

We are also constantly sharing our words and ideas for free. 'Muni' can also be found in 'com*muni*cation'. Just as community is formed through gift-giving, all language and communication can be understood as a gift, made ever more abundant now with the Internet. Academics 'give papers' to contribute their knowledge to the collective.

Some people connect Mardi Gras with the ancient Roman pagan celebrations of spring and fertility. It has been running in New Orleans since the seventeenth century, and a few centuries later inspired Perry Chen to found the creative crowd-funding platform Kickstarter: 'that is where the idea was born – being in the midst of this community of people – neighbors, teachers, artists, mailmen, kids and all generations – no sponsorships, no commercialism, just pure creativity, joy, goodwill and renewal.'

As Lewis Hyde argues in his book on creativity and the gift economy (also titled *The Gift*), artistic talent is in many ways *given* (by genetics, parents, divine intervention or just plain luck) and therefore to be shared. Recall, we call talented people 'gifted'.

Sharing in the digital age

Information gains rather than loses value
through sharing. While the exchange economy
may have been appropriate for the industrial age,
the gift economy is coming back as we enter the
information age.
Gifford Pinchot, 'The Gift Economy' in
Business on a Small Planet

The Web began as a gift, and has spawned a contemporary renaissance of sharing. A few years after Tim Berners-Lee sketched the initial idea in 1989 at CERN, he offered his idea to be available to anyone to use, royalty free, with no patent. He sees the Web as a 'medium for positive change' and in 2009 he launched the World Wide Web Foundation dedicated to its improvement and accessibility.

Continuing in the spirit of his gift, open-source technology whereby code is given away for anyone else to use and develop, has fuelled the growth of the Web. A quarter of all websites are built using Word-Press, a free open-source publishing platform. Much of this works on the principle that you can use the code for free, as long as you share any updates or changes you make, so sharing necessitates more sharing. To date, over a billion works have been licensed through Creative Commons: a tool for artists, musicians and photographers to share their content, for others to use for free.

In a nice meta-experiment, if you look up 'gift economy' on Wikipedia, it cites Wikipedia as one of the biggest gift economies at scale. Wikipedia is an extraordinary success story and testament to the power of gift economies: volunteers around the world contribute their knowledge, and in return get access to the rest of the world's knowledge. It is a non-profit organization, run through the donations of citizens, available in over 300 languages, and has remained the fifth-biggest site in the world since 2015, used by billions of people annually.

In the words of its founder Jimmy Wales, Wikipedia 'makes the Internet not suck'.

'One of the interesting things about Wikipedia is the sheer number of hours that people donate freely of their time: millions of people have contributed to Wikipedia to give this amazing gift to the world,' Jimmy told me. I had met him early on in the process of building my idea for a digital gift economy of skills, and he became an adviser to the project, a board member and a close friend. We both were, and remain, passionate about what he calls the 'third way'.

'Nothing like this works unless it's fun. Normally in society we think mostly about either getting things done through the use of force (i.e. tax) or through trade. There is a whole third way, a sector that doesn't get nearly as much attention which is the gift economy. You know, when you take your grandma to the grocery store, is that charity? No. Is it for money? No. Is it something you do because it's a great thing to do and it makes you happy? Yes.'

In a similar vein, Freecycle has been used by over 9 million people to give away objects. Their mission is to 'build a worldwide sharing movement that reduces waste, saves precious resources & eases the burden on our landfills'. A non-profit, run by volunteers through local Yahoo groups, it is extraordinary how successful the platform has been considering its financial and technological limitations. Freecycle has spawned multiple similar sites, such as Freegle. Meanwhile, other digital platforms such as Gumtree and Craigslist are also used to give away free items, as are local social networks which are beginning to take root.

Millions of people use Couchsurfing to share their beds and sofa space. The website enables people to sleep, for free, in a strangers' spare room or empty sofa. People's motivation for using the platform is sometimes economic – a cheap way to travel the world – but the platform's success ultimately depends upon social motivations: the host wanting to meet new people. My friend Edward occasionally hosts at his flat in San Francisco: he looks for 'interesting' people, and has met a man hiking the whole west coast of South and North America, and a Russian couple who were travelling the world on a hand-made raft!

In her introduction to Lewis Hyde's *The Gift*, Margaret Atwood compares the way that people share content online, through 'free' 'social sites', to the impulse to give: 'Gift exchange demands reciprocity and is fed by it: thus, one Retweet deserves another, and those who offer advice for nothing may expect to receive it for nothing when in need.' Given that many of these social sites are not really free (as discussed in Chapter Four), you might read this as a distortion of the principle.

All these developments in technology reflect a deeper cultural shift. In a 2016 study of millennials from around the world, 86 per cent defined wealth as something other than having possessions or cash. Having meaningful 'experiences' was the most popular life goal in eight of the ten countries surveyed. Meanwhile, services like Netflix and Spotify have made owning content fairly redundant: why own ten albums, when you can have access to millions? As people come to realize that time is the most valuable asset we have, we are transitioning into an age where access and experience are more prized than ownership.

With my research banked, I left university convinced that a bigger, more organized gift economy was both possible and positive. There seemed to be a space missing for facilitating a gift economy in favours and time. Having just written my thesis on the need to redefine utopia and debunk what we consider impossible, I decided to call this new project Impossible.

Attempting the Impossible

> A system of exchange – as an alternative to the
> monetary system – based on a currency of
> reciprocal good will – as made transparent
> (visually represented & facilitated) by the website.
>
> My diary, 2011

Impossible was built by its own ideas. So many people loved the idea, and got on board, that I soon felt the responsibility of a captain sailing a ship that had already left the port.

We decided to design Impossible as a social network because that felt like a natural home for building community. I had never been particularly adept at using software tools, so I made a few attempts at hiring people to create '<u>wireframes</u>' for me. Yet every time they failed to realize the images I had in my head.

I started to get frustrated, then I had a breakthrough moment: why don't I just collage them?! Late one Sunday I stayed up through the night, cutting and glueing together a design for the website. But there was one thing I couldn't figure out: what do you replace money with?

Conceptually I was clear in my mind that I didn't want to replace it with money by a different name i.e. an alternative digital coin or token system. There were already plenty of experiments in that space, from TimeBanks, to local currencies, or cryptocurrencies. Whilst these initiatives offered interesting

Both a style of lightweight eyewear and the name for the illustrated blueprint of a digital service.

311

developments, they weren't seeking to move away from an exchange system. I was looking to design a gift-based system that would release all those happy chemicals and make social bonds.

That said, I could see that it would be really difficult to establish trust and mutual reciprocity in this context. Research suggests that the human brain can manage to remember up to 150 people, and then it gets hard. How do you build reputation and trust in giant mega-cities of millions of people? I considered a rating system – like eBay – as a way to establish reputation, but it hardly felt appropriate to rate someone after they had done you a favour.

Through a chain of mutual friends, I met with the ingenious Tea Uglow who was running Google's Creative Lab, and became a turning point of inspiration for the project. Tea and I were sitting in the Google café in London – essentially a posh kindergarten – and I was telling her about the idea. I talked about the currency of gratitude I was trying to establish. At this point I was thinking about an abundant currency: one that you could only ever earn but not spend; that allowed users to see who was more active and generous on the network, and thereby potentially reward them. Tea's mind buzzed around in circles at a hundred miles an hour, reaching out to the cosmos and then back again. 'Well, why don't you call it Thanks?' she ventured. And there, like that, so simply, was our solution. A Thanks currency.

Who Cares Wins?

My grandmother always enjoined our family
to be forever practising the discipline of
gratitude. Impossible really is the giving tree,
in practice, on a global scale and with constant
regeneration.
Chelsea Clinton, at the US launch
of Impossible

The other central concept in the platform we developed was that of wishing. I had long been
drawn to the idea as a universal language: around
the world are trees, ponds and walls that enshrine
the idea of a wish. I first encountered a wishing
tree at a temple in a park in Tokyo: you write your
wish on a wooden plaque and hang it on the tree
for the cosmos to fulfil. I wanted to take this idea,
and ground it: by showing your wish to your community, it becomes a conversation, and it might just
come true.

The other challenge to design was the legal structure to sit Impossible in: making it a charity felt
instinctively cumbersome and disempowered, but a
business wasn't appropriate. In researching the possibilities for 'social business' I happened to meet
Professor Muhammad Yunus – the ex-economics
teacher who won the Nobel Peace Prize in 2006 for
micro-finance. He is now dedicated to championing
non-dividend social businesses that are set up to solve
problems, and reinvest 100 per cent of profits back
into their purpose.

We travelled to Bangladesh, where a group of other enthusiasts and I followed Professor Yunus through the villages with wicker huts designated 'banks'; and examples of social businesses that his work had inspired. I left clear about why I had come: gifted with a vision of capitalism that could be evolved to become an agent of positive change. When we returned to London, with the generous help of our lawyers we made Impossible the first 'Yunus Social Business' in the UK.

> The way we live our lives, we become islands. We
> lose contact with each other. In a gift economy,
> connecting people, islands come closer to
> become a bigger island.
> Professor Muhammad Yunus, Davos, 2013

When I and our team of friends, helpers and employees launched the platform in 2013, we reached over 100 million people with our message. Alongside media, we erected wishing trees around the world: the giant, beautiful, ancient chestnut tree in front of the chapel at my old college, King's College, Cambridge, was decorated with students' wishes; lines of trees inside the Science Museum in London; we installed a wishing tree in a New York art gallery; and finally hung wishes on a beautiful banyan tree in front of Sydney Harbour. Meanwhile a roboticist friend in San Francisco recycled an ATM, painted it white, and plugged it into our app so you could print wishes!

Who Cares Wins?

> You give but little when you give of your
> possessions. It is when you give of yourself
> that you truly give.
>
> Kahlil Gibran, *The Prophet*

Over the course of the next year, Impossible managed to attract more than 100,000 users in 120 countries worldwide. Meaningful connections were made between individuals who would never otherwise have found each other, and I received more messages and emails from users to thank me for the people they had met through the platform, than for the gifts they received. The community was eclectic and diverse, with active members from Ghana and Australia, to the US and UK. The ages ranged from teenage to octogenarian, and it was roughly equal in terms of gender split.

Personally, I received cooking, piano, driving and French lessons, and even hosted a train enthusiast from Scotland to help me build a model train set. I let people borrow my flat when I was away, gave advice on sustainable fashion, provided the voice for an arthouse film and played the role of witness to two Italian women's wedding in Cambridge.

We developed an editorial platform capturing some of the stories: a heartbroken girl who had lost her camera was sent one from another sympathetic user; a hard-up couple got a free haircut before their wedding day; a tent was donated for a homeless man in winter; Christmas and Thanksgiving lunches were hosted for strangers in others' homes. Small, simple and human gestures. Most of the actions through Impossible were difficult to trace and I would just catch ripples of what had happened when I met a member who would say anecdotally, oh, yes, I have been giving English or piano lessons through Impossible for years.

There was also an emotional element to the wishes that I hadn't anticipated. One of our most popular posts, for example, was wishing for the war in Ukraine to end, whilst others wished for family members to get well. Of course, these weren't actionable posts, but they created empathy and communal storytelling. Sometimes someone just needed to feel heard, receive advice or have their pain acknowledged.

Our biggest community was in London where we held regular events to bring people together. Usually they would take the form of a workshop with someone sharing their skills with a group of people: such as Super 8 film-making, or choral singing. Paul McCartney heard about what we were doing and offered a free music workshop to a group of thirty musicians, and my writer-director friend Joss Whedon gave an impromptu screenwriting session. We recorded some of these workshops and posted the content online and in a magazine, *Impossible to Print*. Impossible became very popular in Brazil and our tiny team had to hustle to add language filters to the platform. A postal gift economy developed where people posted letters and physical – often handmade – gifts to one another.

Looking back, there was a real magic to these moments. Impossible wasn't achieving the show-stopping, world-changing, level of success I had been hoping for – it wasn't making a dent in the global economy – but it was demonstrating that the premise was correct. People were showing with their words, actions and especially their time, that they wanted to be part of a community. That kindness is a currency of sorts. In spite of numerous technical issues and design challenges, our users continued to come back and look for opportunities to give.

Impossible takes time

There was a problem. The moneyless platform was (paradoxically) expensive to run and since we had designed it to be entirely non-profit, we couldn't easily raise investment. I was self-funding it, with help from an 'Innovation in Giving' grant and generous gifts of people's time but I soon learned that technology is an expensive medium to work in. I continued to volunteer my time, but we had full-time engineers and staff to pay, without a clear business model in sight. (I wasn't keen on advertising for all the reasons discussed in Chapter Four.)

The platform I had built, focused on a manifesto about the precious value of time, was also ironically taking all my time. Overworked, and aware of the pickle I'd found myself in, I made a Christmas card that year with a drawing of myself on the computer, saying, 'Sorry, too busy to speak right now – trying to make the world more social!'

'You can't change the system from inside the system!' a middle-aged musician I admired for his politics angrily shouted at me early on, enraged that I had asked his opinion on my new gift economy idea. Those words haunted me throughout – more so than the many words of encouragement I received from most others.

He wasn't right to shout but—

Was he *right*? Perhaps.

> The Master's Tools Will Never Dismantle the
> Master's House.
>
> Audre Lorde

A year after launch, I had a photo-shoot in Hong Kong and arrived to that concrete sci-fi reality feeling burnt out and unwell. I went swimming in the hotel pool, floating between a thousand dark windows and spinning air-con fans, feeling thoroughly disillusioned by life and capitalism. Back in my hotel room, I was hit with a wave of the most intense food poisoning I've ever experienced. I proceeded to projectile-vomit all over the floor and scrambled to find a phone to call for a doctor. A few hours later, when I was asked to pay the visiting night doctor, I searched the room for my money, only to realize that my credit card was in the back pocket of my denim shorts, pulled off in haste and now lying amidst yesterday's lunch. The fact that I had thrown up all over my credit card felt somehow poignant.

From Hong Kong, I travelled to Ladakh, India, where my partner and I trekked through the very remote Zanskar valley – a pass that runs between the 6,000-metre-high Zanskar mountain ranges and is only accessible by foot. After a few days of trekking, we were deep in the valley, sleeping in a local's house. I woke up very early, to the sound of beetles scratching in the wall behind my head, and slipped quietly out of bed to walk into the bright green barley fields. There I found the wooden trunk of a tree, lying on its side; I sat on it and watered the barley with my tears until I had run out. I had put as much money and time as I could personally afford into Impossible, and it had nearly bankrupted me. I felt like the project was failing and therefore I was failing.

As we left Ladakh, we drove across into Kashmir and passed a yellow bridge. India has this charming habit of writing poetic lines across road signs, and this one caused me to park and jump out of the car. Underneath the yellow bridge it said, 'Difficult will be done immediately. Impossible takes time.'

After struggling to find a way to financially sustain the gifting platform, in 2017 we stopped investing in it, and released the code we had built under an open source licence. It has since been used by several communities – from a creative community in London to share favours, to a refugee organization seeking to connect grass-roots organizations with volunteers. We pivoted Impossible as an organization to start working on other products and services under the guiding principle of 'planet-centric design'; and published the

Since the mid-twentieth century, businesses have been creating products with the customer at the centre of the creative process: 'user-centric design' has created a short-term convenience culture. Press a button and get a taxi. Press another and get food delivered in thirty minutes ... Planet centric design is a methodology that puts the planet at the centre of the creative process and in doing so creates products that help meet global ambitions (seventeen UN goals), not just individual ones.

'Impossible People' code – to be picked up by the next crazy adventurer who wants to use technology to foster a gift economy.

Meanwhile, the Yunus Social Business (YSB) proved a difficult model for us in practice. We wanted to raise investment, but most technology investors – even 'impact' investors – expect a financial return. There are <u>over 1,500 YSBs</u> in operation globally, and I still believe that the absolute commitment of YSBs to return 100 per cent of their profit to their purpose offers the deep paradigm shift in capitalism that we ultimately need: removing the opportunity for extreme inequality that capitalism creates, whilst mobilizing its power and efficiency.

One particularly impressive example of a Yunus Social Business is 'Make Sense', which helps incubate social enterprises around the world.

A more mainstream version of social business was gaining momentum and as soon as they launched the structure in the UK, we transitioned Impossible to be a certified B Corp. B Corps allow for dividends, but like YSBs, they require a fundamental change in the legal underpinnings of certified companies, to make them legally accountable not just to shareholders, but to customers, employees, the environment. B Lab is a non-profit that audits and certifies the companies according to broad ethical criteria.

Meanwhile the 'sharing economy' – all those niche 'collaborative consumption' sites I had been studying in my Cambridge days – exploded into the mainstream. Airbnb became a global success story, allowing people to open up their houses or spare rooms to strangers at rates typically cheaper than hotels. BlaBlaCars helps millions of people to car-pool. Lyft allows people to rent out their own cars as personal drivers, and Uber 'pool' functionality to turn cars into small buses.

Arguably more rental platforms, or 'gig economy', than sharing: some early evangelists are disappointed that they have usurped the potential for real sharing, and created other negative side-effects. Yet maybe they do foster new peer-to-peer ways to connect with others, and encourage us to open up our spaces and resources for others to use.

The hard-won truth

Maybe it was always obvious: you can't build a gift economy in a capitalist framework. Perhaps the gift economy needn't be structured or organized at all. Perhaps I had misunderstood the metaphor of the birds in a V: yes, they (we) will fly together, fairly and generously, but only in an organic, decentralized, natural way.

Indeed, the most successful, large-scale examples of giving through Impossible were elicited in a spontaneous, messy, self-organized way, through communication not technology. From the beginning, as I started talking to people about the idea they would say, 'I love it. How can I help you?'

Thus, Impossible was created by the ideals it was conceived to promote. The gift economy wasn't the app – it was a spirit to be tapped into. And the journey was proof of concept: most people like to give, are inherently kind, and want to be part of a kinder world.

The hard-won truth is that you don't need a fancy piece of technology to enable community. It is created through sharing, kindness, and realizing our mutual interdependence. There are no barriers to entry. The strongest communities I have witnessed are in areas that are both financially and technologically poor, like the Burmese refugees, or the men I encountered in northern Ghana who circled around a mud hut, working together to fix the thatching.

Perhaps the most profound thing I learned through my experience of trying to enable an online gift economy, was how to receive. Working on Impossible, I came to realize that a gift economy should really be called a 'giving and receiving culture' because for every gift there is a recipient. Receiving is not always straightforward or comfortable. In some ways, it is easier to be a giver because then you don't have to feel any obligation to other people, whereas to receive requires you to be vulnerable. Yet to receive is also to give, because you grant someone else the opportunity to give.

The gift economy also brought me the biggest gift: my daughter, Wylde. I met her father when he offered to help develop the technology for Impossible as a gift. When I introduced him to Professor Yunus, Yunus's face lit up and he exclaimed, 'Ah – Mr Impossible!' Was it prophetic or fated?

Who cares (who) wins?

Impossible hadn't manifested in the form I had wished for, of one big sharing social network, yet my original mission to develop a meaningful community was achieved. When I got to my thirtieth birthday, seven years after first having the idea for Impossible, I realized that I had met half of the guests through it. The idea may not have changed the world, but it changed me.

I learned a lot through the process, and am still learning, but I feel very certain of the reality, power and pervasiveness of kindness. Of the possibility we have to reclaim the only things of real value: time, experiences, relationships. That this process of rec-lamation might make us happier, and our societies more sustainable. And that this process – towards community finds its roots and fruits in immediate and simple ways.

To embody more resilient, connected communi-ties, what we need – more than any new structures, apps, communes or policies – is a shift in our mind-sets: to fight the idea that we are competing, and instead to share, to accept, to trust.

Humans are a naturally social species. We came this far, through savannahs, and across oceans and forests, for hundreds of thousands of years, by work-ing together: by cooperating.

Beyond the social and psychological benefits, strong community may also be our greatest tool in confronting our myriad challenges, including the climate crisis, as we recognize ourselves as part of a global community, of billions of people sharing this precious planet. If we can learn anything from indigenous communities, it is that the ability to share better with one another will be key to our long-term survival.

Chapter Twelve
Modern Fairy Tales

Slow growth, UBI and simplicity

> There are only two things you can spend
> in this world. Money. And Time.
> Right now you have more
> Time than you'll ever have.
> You're richer than you'll ever be.
> Right this second. This minute. This day.
> So how are you going to spend it?
> Do something impossible.
>
> Jez Butterworth and me,
> The Impossible Manifesto

Our attempt at the Impossible was in many ways about reclaiming the only thing we ever really own: time. Yet something that quickly became self-evident is that *time* is a luxury that many people simply cannot afford to give. People could manage an hour here or there, but generally we are very time poor.

Our societies are working, on average, two to three times longer hours than our gatherer-hunter ancestors. The gadgets and devices that promised to save us time – washing machines, cars, phones – are expensive to maintain; and many of us have been caught up working hard in Charlie Chaplin's wheels of *Modern Times*.

When the world is wealthier than it's ever been, and when time is such a precious resource, why are we so poor in it? Where did we go wrong? Can we get some of it back? Could working less save the planet, and our sanity alongside?

Chasing growth

Only in economics is endless expansion seen as a
virtue. In biology it is called cancer.
David Pilling, *The Growth Delusion*

I woke from a dream. In it, my daughter and I had been playing in the garden and she was asking me to learn, from the flowers and plants that surrounded us, about growth. To think about slow growth, or even de-growth. How nature takes its time, but still gets everything done.

Ecologists have long argued that our obsession with economic growth is central to our erosion of the planet. President JFK's environmental advisor, Kenneth Boulding, famously quipped, 'Anyone who believes exponential growth can go on forever in a finite world is either a madman or an economist.'

If we can map economic growth, alongside carbon emissions, they are historically wedded together.

The impact of 'growth' applies on both a government and a personal level. While researching this book I came across a rather disturbing study – *Good Intents Low Impacts* – which analysed the environmental intentions, environmental impacts, and income levels of over a thousand people in Germany. It found (as other studies also had) that income was a more predictable indicator of someone's environmental impact than good environmental awareness.

The logic goes: environmentally aware people may be more likely to eat less meat, and buy organic food or eco-appliances, but if wealthier, they are still likely to have a bigger carbon footprint because of the high-impact 'consumption options that open up with higher socio-economic status' like big houses, flying and driving. In fact, being wealthier was – paradoxically – likely to increase *both* environmental awareness *and* environmental impact.

The research correlates with other statistics: we live in a world where the richest 10 per cent are responsible for half of all emissions.

Meanwhile, countries, companies and many individuals continue to aspire to economic growth. There is often a tension between a vision of philanthropy that seeks to lift everyone – 7.5 billion people – globally towards a Western lifestyle, and the reality that the typical Western lifestyle is not even sustainable for the minority who are currently enjoying (or not necessarily enjoying) it.

As we continue to encourage the global community to step up into development – as we should we also need to wonder, what future are we developing?

Are relatively affluent, environmentally conscious people (like myself) completely delusional?

Do we need to be more honest with ourselves about what sustainability really means?

Or can we have our capitalist cake and eat it?

Some people remain hopeful we can.

Whilst on the topic of simplicity versus capitalist cakes, I'd like to give a nod to honey – or rather the documentary *Honeyland* – a beautiful environmental parable, about a bee keeper in Macedonia.

327

Consciously uncoupling: break-ups in a fairy tale

> All you can talk about is money and fairy tales of
> eternal economic growth.
> Greta Thunberg, the UN Climate Action
> Summit, 2019

There has been an emerging phenomenon in recent decades in some 'higher-income' countries (the UK, France, Germany, Sweden and the US) where emissions have started to de-couple from economic growth. The marriage lines are breaking apart: scissoring open. This is not yet true of 'lower-income' nations, but they seem to be heading in this direction, given the declines in their 'emissions intensity' (the ratio of greenhouse emissions produced to GDP).

Critics argue that these calculations are guilty of 'creative carbon accounting' – they often don't consider factors like international aviation, shipping, or products manufactured abroad. Yet it does offer hope, that as we decarbonize our economies (through better energy efficiency, and carbon-free energy) we can decouple growth from ecological disaster.

Indeed, in *Enlightenment Now*, Steven Pinker indicates another study which has found that 'people with stronger emancipative values – tolerance, equality, freedom of thought and speech – which tend to go with affluence and education, are also more likely to recycle and to pressure governments and businesses into protecting the environment'. So the logic goes, wealthier eco-people might emit more, but they will *also* push for systemic changes.

'The long sweep of decarbonization shows that economic growth is not synonymous with burning carbon. Some optimists believe that if the trend is allowed to evolve into its next phase . . . the climate will have a soft landing. But only the sunniest believe this will happen by itself . . . Instead, decarbonization needs to be helped along with pushes from policy and technology,' Pinker comments, who advocates carbon pricing, technology (especially nuclear energy) and afforestation/reforestation as the main tools for driving deep decarbonization alongside economic growth.

The fairy tale may yet prevail.

Slow Growth

Whilst pursuing deep decarbonization, is there another part to the narrative we might also want to consider?

In the face of climate crisis, might it be worth reflecting on our obsessive attachment to growth itself?

To pause to wonder what in fact it is that we are trying to grow?

Could a strategy of slow growth (or even degrowth) help better ensure our survival?

Might it actually make us happier?

Economic growth, centred around Gross Domestic Product (GDP), is almost a religion in modern times, provoking calls of heresy when questioned. It is the headline promise of most political candidates, and drives the focus of our companies, and often our personal sense of achievement.

In *The Growth Delusion* David Pilling argues that GDP does have value as an imperfect indicator of progress; but we need to recognize its limitations rather than enthrone it as our god. Contemporary measurements of GDP do not, for example, measure the value added by free public services like education, volunteering or housework; or take account of environmental degradation; but they do include factors that fail to contribute to human well-being such as crime or fuel spent in traffic jams.

Indeed, Simon Kuznets, who first developed the modern concept of GDP in 1934, warned against its limitations as a measurement of human welfare. He proposed a more nuanced approach that deducted income from things which have a negative effect on society, like armaments: 'Distinctions must be kept in mind between quantity and quality of growth . . . Goals for more growth should specify more growth of what and for what.'

Pilling advocates contextualizing GDP alongside other metrics, such as the 'wealth' of nations. Akin to how we might measure our wealth according to what we own, as much as what we earn; a wealth of nations would reflect not just economic activity (as GDP does) but also their cultural and natural assets. Right now, if a country were to slash all its forests to produce beef and soya for sale, or mine and sell its gold, that might reflect well in terms of GDP but poorly in terms of actual wealth.

Economists and scientists have also long advocated 'green GDP' which factors in the environmental impact of growth. An estimate of this in China found the result to be a third less than the official GDP, once you take account of waste, social costs and environmental damage. The Dutch government are taking the lead: bringing together their finance and climate ministries, and committed to having a fully circular economy by 2050.

The very purpose of life is to be happy.
Dalai Lama

Finally, can we measure that most evanescent thing: happiness? Numerous studies suggest a disconnect between infinite economic growth and infinite happiness: finding that after people earn a certain amount, emotional well-being plateaus. Meanwhile, the New Economics Foundation's research using surveys on happiness (the Happy Planet Index) found that whilst the entire European continent's carbon footprint has increased by 70 per cent since 1961, life expectancy has grown 8 per cent but happiness levels have not risen.

Bhutan famously tries to prioritize well-being by measuring Gross National Happiness. After realizing that a third of families include someone who is mentally ill – and 10 per cent of adults experience depression – the UK government have started to track

Bhutan's Gross National Happiness index was first introduced in 1972 by King Jigme Singye Wangchuck, the 4th King of Bhutan. It measures nine key criteria: Psychological well-being, health, education, time use, cultural diversity and resilience, good governance, community vitality, ecological diversity and resilience, and living standards.

national well-being, and New Zealand has become the
first country in the world to measure success by their
citizen's well-being, through running a 'well-being
budget'.

How working less could save the planet
(and our sanity)

During the 1930s economic crisis and Great Depression, in an essay entitled 'Economic Possibilities for
our Grandchildren', John Maynard Keynes speculated
that our technological progress and the accumula-
tion of capital would over time gradually free more
and more people from the fundamental 'struggle for
Receptionists,   subsistence' that had plagued humans, animals and
corporate lawyers,   insects alike since the 'beginnings of life in its most
telemarketers,   primitive forms'.
public relations   Keynes speculated that, in part because of the 'new
specialists, admin- disease' of 'technological unemployment', by 2030
istrative assistants most people would stop 'blindly pursuing wealth' and
(and no doubt   be working perhaps 'three-hour shifts or a fifteen-
fashion models   hour week'.
too) all fall under   Yet even though automation has been displacing
the category of   jobs for decades, many of us are not working any
'bullshit jobs' as   less. Instead, we seem to have created ever new and
defined by anthro-   shifting types of work – what anthropologist David
pologist David   Graeber calls 'bullshit jobs' – largely to fuel GDP and
Graeber in his 2018 consumption.
book of the same   There is a beckoning counter-culture push against
name. In fact,   this trend. Whilst environmentalists across the board
Graeber asserts
that over half of
all societal work is
pointless.

call on people to reduce consumption, many others have long advocated a decreased working week as part of what is called 'slow growth', 'de-growth' or 'steady-state economics'.

To be clear, 'work' in this context is defined as doing an activity that you wouldn't otherwise want to do unpaid. The focus of this inquiry also pertains to high-income countries, mostly in the Global North, where consumption rates are significantly higher, and most infrastructure has been developed already.

> Many scholars have argued that continued economic growth in the global North is antithetical to achieving global environmental sustainability. An increasingly prominent idea is that developed countries could achieve slower or zero economic growth in a socially sustainable way by reducing work hours . . . While the 1970s shift to focus on productivity, profits and growth made sense at the time, we believe that a new conversation about growth in wealthy countries is long overdue.
>
> Kyle Knight, Eugene A. Rosa and Juliet B. Schor, 'Reducing Growth to Achieve Environmental Sustainability: The Role of Work Hours', working paper no. 304, Political Economy Research Institute, University of Massachusetts, Amherst

By analysing the possible outcomes of a reduction in the average working week, one study in Sweden found that working hours and emissions are tied together: a 10 per cent reduction in working hours would see an 8 per cent reduction in emissions. A study from the US estimated that reducing the average work week by 0.5 per cent every year over the course of the rest of the twenty-first century, would mitigate a quarter to half of global warming not already 'locked-in' by 2100 as a consequence of an overall reduction in production and consumption.

Nor is this a miserable exchange of prospects. In fact, studies suggest that free time – over and above income – is a key ingredient for happiness. People who are more abundant in time are more likely to engage in low-consumption activities: like spending time with family and friends, gardening, being in nature, having hobbies, learning new skills, volunteering – doing impossible things.

When the data on working practices and impacts in Europe and the US was compared, it transpired that people who live in countries which work fewer hours, have a lower environmental footprint and are happier.

So, if we and our governments start chasing Gross National Happiness as readily as we do Gross Domestic Product, we might find that planetary and mental health track together.

Our wild, precious life

What would you like to do if money were no
object? How would you really enjoy spending
your life? It's so important to consider the
question: What do I desire?
Alan Watts, 'What if Money was No
Object', YouTube video

If you didn't *have* to work so much, or for money,
what would you do? Would you do a 'job' you love?
Spend more time in a hobby or leisure? Would you
give more of your time to family? Is it possible to imagine a world where unemployment is the norm, but
social unrest isn't?

What would less (material) growth look like, and
what other types of growth (spiritual, social, psychological, environmental) might find space to emerge
instead?

A 'silent revolution' has arguably happened in the
West in recent decades. As an increasing number of
people have been able to meet their survival needs,
we have entered what some call a 'post-material age':
witnessing the rise of values such as self-autonomy,
self-expression, freedom of speech, gender equality
and environmentalism.

Indeed, minimalism is often a side-effect of affluence: an over-exposure to stuff can make people
appreciative of space and simplicity. Having been
poor, wealthy and places in between, this was certainly my experience.

My sister and I were brought up in London by our mother, who had a very low income. Then I began my professional life – aged fourteen – as a face of consumerism. The catwalk catapulted me into opportunity and privilege. I've been in adverts, posters and paparazzi shots. In fact, I have even appeared in a North Korean propaganda film criticizing illusion and deception in the Western world! My face appears, blinking long false eyelashes, with the label 'ILLU-SION' printed on top.

Etymology: from Old French, literally 'dead pledge', from *mort* (from Latin *mortuus* 'dead') + *gage* ('pledge').

I got lucky and remain grateful for all the new experiences and freedoms, but over time I came to understand my lifestyle to be a trap – making me a slave to a work cycle, ever anxious about money, especially under the 'death pledge' of a <u>mortgage</u>.

My financial anxiety wasn't helped by the fact that I stepped away from modelling quite quickly, became picky about who I would work with, and started volunteering a lot of my time. But I had stuck my head in the belly of the beast, and I was keen to pull it back out. As my bank account shrank, I found myself rich in new ways – in experiences, relationships, and doing work I believed in.

Now, I appreciate minimalism, and having less. I still live a privileged life but over the last few years I have sold or given away many of my things, and try to be mindful about only buying what I need or *really* love. I am on a journey towards simplicity, though far from arriving.

Voluntary simplicity

I have spent my days stringing and unstringing
my instrument while the song I came to sing
remains unsung.
Rabindranath Tagore, *Gitanjali*

Lots of people around the world, like my friend Mark Boyle, are going 'back to the land' to re-find the simple life. In Northern California, I visited Salmon Creek Farm one spring weekend for a community potlatch and found a collection of interesting people, good food, beautiful trees and plenty of '<u>cabin porn</u>'.

Salmon Creek Farm (SCF) is a thirty-three-acre redwood forest populated by eleven quirky wooden cabins originally hand-built in 1971 by a group of eleven friends. They, like many others at the time, were exhausted by the mainstream political culture – the Vietnam War and mass consumption – and escaped to raise their children in a simpler and more self reliant way. According to co-founder River, they were looking for 'a new pattern of living that does not rip off the planet or any of her inhabitants'.

In 2014, artist Fritz Haeg went looking for a place where he could build 'community' and stumbled on SCF. Since then he has been inviting an extended creative and queer community to come and spend time there, participating in the building and running of the space, and seeing how the landscape affects their artistic practice. All food is grown on site or bought from local farms; Fritz aims for a policy of 'zero waste'

Not quite what it sounds like, or maybe exactly what it sounds like, depending on what way your imagination slants. An incredibly popular lifestyle Tumblr turned incredibly popular coffee table book showing pictures of trendy cabins in pristine locations.

(meaning no packaging). Visitors are encouraged to make a $10 donation when they stay, and invited to peruse the abundance of books and magazines.

The forest breathed into my bedroom through large glass windows as I read *Woman and Country* magazines, then proceeded to dream that I was infected with greed: an evil spirit that took over my being, and pushed me to lose empathy with the world. That night, back in San Francisco, I watched the Japanese animation *Spirited Away* with my daughter. 'Why did the parents eat too much and turn into pigs?' she kept asking me. 'I guess because they were greedy,' was the only thing I could think to say.

> He who knows he has enough is rich.
>
> Lao Tzu

As simple as the name implies. A beautiful book by Duane Elgin, and general life philosophy: 'outwardly simple, inwardly rich'.

Igniting a feeling of great pleasure, like a miniature firework. The correct order for tidying, according to Marie Kondo, is as follows: clothes first, then books, then papers, then *komono* (miscellaneous items) and finally sentimental items.

There is a burgeoning counter-culture resisting the mainstream insistence that we have infinite, endlessly shifting needs that are impossible to meet. An increasing number of individuals and households are opting for voluntary simplicity: to clarify their lives and lower their costs. It doesn't have to be all hemp and hippy: sometimes minimalism, clean lines, quality objects and focused actions, can look and feel really sexy.

Marie Kondo, the Japanese 'organizing consultant' and author, has sold millions of books and launched a Netflix show encouraging people to de-clutter their lives, and practise gratitude to objects – rooted in Shinto Japanese philosophy. Kondo recommends only keeping objects that 'spark joy' and letting go of those that don't – not before thanking them for their service.

It is hard – if not impossible – to speak of simplicity, without also addressing equality. The studies advocating a reduced working week acknowledge that this would be difficult to achieve in countries where inequality is high and/or growing, without problematically compromising many people's living standards. Many people simply cannot afford to work less.

As French economist Thomas Piketty has put it, the dominant model of capitalism operates as a 'fundamental force for divergence'. While 10 per cent of the world's people are still living in 'extreme poverty', struggling to meet their basic survival needs, in 'wealthy' nations, the income gap has widened so much in recent decades that many people are struggling to stay afloat even as they work two, three or even four jobs.

In the US since 1980, whilst GDP grew by 154 per cent, the median household income grew just 16.5 per cent. Public and private wealth have also diverged: since the 1970s, whilst private wealth in the UK has more than doubled (from 300 per cent to over 600 per cent of GDP), public wealth (assets minus debts) has diminished from about 50 per cent of GDP, to a negative value today.

So is the fifteen-hour working week a pipe dream? Or is there a way to fund the shift?

A lot needs to be done from a political perspective to see that the overall wealth we have created is shared. Even corporations, and wealthy business leaders, recognize this fact. In 2019, nineteen US billionaires

The Accountable Capitalism Act is a law proposed by US Senator Elizabeth Warren that seeks to make companies with revenue over $1 billion legally responsible to all stakeholders (shareholders, employees, the environment) – making mandatory what social businesses have done voluntarily.

signed an open letter calling on a wealth tax. The drive for more 'accountable capitalism' would also certainly help.

Finally, there is Universal Basic Income (UBI) – an old idea whose time seems to have come.

UBI made its first appearance in 1516 in Thomas More's aptly titled *Utopia*, and is now grabbing headlines and political attention around the world. It claims to reduce poverty, improve welfare, deal with the inevitable displacement of jobs brought about by automation, bring greater social equity to society, and free it of the 'bullshit jobs' so demanding of time, happiness and consumption of planetary resources.

What do we mean when we talk about Universal Basic Income? Is it economics, altruism, politics? It's an apparently simple solution but is it viable? Is UBI a reason to be optimistic?

Like the Green New Deal, UBI manifests in different ways in different places, but the underlying premise is generally quite simple: give absolutely everyone a minimal basic income that will act as a universal security net.

> Basic income, even a modest amount, would give us a sense of basic security [which is] a human need. If everyone had a basic income security, it would increase mental bandwidth, and increase tolerance and altruism towards the other. For me, it has to be the answer for the breakdown of our income distribution system.
>
> Guy Standing, *Basic Income*

Some pilots of the idea are being funded by philan-thropists, but there have been alternative proposals for paying for it long-term: perhaps it could be funded through meaningful carbon pricing, combining two transformative ideas in one.

'Carbon dividends represent a strong moral and politically practical justification for basic income,' argues long-time UBI proponent Guy Standing. Others suggest funding it through national assets, like a Citizen's Wealth Fund, that is run separate from government. Most often it's proposed as a revision to some elements of means-tested welfare spending.

A public wealth fund owned by the nation's people (separate to gov-ernment, business or private interest). Like the Alaska Permanent Fund.

A large part of existing welfare is spent policing and measuring the 'deserving' poor: an expensive and often inhumane process that makes the most vulnerable members of our society jump through Kafka-esque hoops to prove their merit; to distin-guish them from the relatively small number of benefit fraudsters. Growing up, I saw first-hand how the system cut both ways: school friends trying to game the system; whilst my mother was dragged through years of paperwork to *prove* her ill health. (Ironically, it's much easier to jump through hoops if you are fit.)

> We could replace the ragbag of specific welfare
> programs with a single comprehensive program
> of income supplements in cash – which would
> do more efficiently and humanely what
> our present welfare system does so inefficiently
> and inhumanely.
> Milton Friedman, *Capitalism and Freedom*

One of the world's most famous examples of a UBI experiment has been led by the centre-right government of Finland, looking to reduce its social security spending, which was 31 per cent of GDP. The UBI pilot was run in 2017 with two thousand unemployed people between the ages of twenty-five and fifty-eight being paid €560 every month. There was no obligation for recipients to seek employment, and finding employment would not affect their benefits.

Sini Marttinen, a thirty-five-year-old consultant who took part in the experiment, said the UBI payments had given her the confidence to start her own business, whilst initial results from the Finnish trial showed it improved the mental health of participants, decreasing stress.

The Finnish trial showed no overall effect on employment, yet fears that UBI would breed apathy are contradicted by one of the oldest UBI schemes in the world: the Alaska Permanent Fund which sees state-owned oil revenues redistributed annually to its citizens. Ranging from $800 to over $2,000 per person, depending on the year, it is not enough for people to live off, but mitigates the capacity for any desperate ($2 a day) levels of poverty. A study that looked to see if the programme affected levels of employment found that it had a 'marginally positive employment effect' by increasing people's engagement in part-time work.

The capacity of UBI to eliminate extreme poverty, as evidenced by Alaska, is important. Just before he was assassinated, Martin Luther King wrote the book *Where Do We Go From Here: Chaos or Community*, in

which he advocated a universal basic income: 'I am
now convinced that the solution to poverty is to abolish it directly by a now widely discussed measure: the
guaranteed income.' Luther King quoted an estimate
by John Kenneth Galbraith that the cost of such a
proposal would be $20 billion – equivalent to what
was being spent on the Vietnam War annually at the
time.

'Wow, what happened to this idea?' asked Michael
Tubbs when he read King's book at Stanford University. The youngest, and first-ever African-American,
Mayor of Stockton, California, Tubbs quickly got
to work once in office. He launched the first city-
sized pilot of UBI in the US: giving 125 families
$500 a month, no strings attached, with researchers
tracking the different impacts it had on community
engagement, physical and mental health and income
volatility.

Stockton is an apt place to stage the pilot. It is
close to Silicon Valley, a hotbed of UBI enthusiasts
and work-displacing automation. And when Tubbs
became mayor, the town had recently declared bank-
ruptcy, one in four people lived below the poverty line,
and it had one of the highest homicide rates in the US.

'The violence that we see in these communities
is a symptom of deeper issues. We're really focusing
on addressing structural issues, and not just allowing
the narrative of the community to be one of crime or
violence or poverty,' said Tubbs. 'The root of issues
of poverty and economic insecurity isn't people, but
systems.'

Robot: friend or foe?

Machinery must work for us in coal mines.
Oscar Wilde, *The Soul of Man Under*
Socialism

In the nineteenth century, Oscar Wilde enthused about intelligent machines that would be 'the property of all' and 'clean the streets, and run messages on wet days, and do anything that is tedious or distressing'. That day is arriving; we already have phones for messages and machines to clean streets, and robots are estimated to replace up to half of jobs in the next twenty years – albeit they aren't owned by everyone.

In 2017, the EU published a draft report on automation from which it concluded 'in the light of the possible effects on the labour market of robotics and Artificial Intelligence, a general basic income should be seriously considered, and invites all Member States to do so.'

Plenty of anxiety-inducing, dystopian articles have been written about robots taking over, but advances in AI could in fact offer a real opportunity for humanity: pushing humans to focus on more specialized, higher-skilled, higher-paid and creative tasks. If the social risks are managed with innovative and sensitive policies, like UBI, robots might enable us to embody Keynes's old dream.

Hyper-aware of the risks of automation, and with an appetite for the innovative, UBI has found lots of ardour amongst Silicon Valley technologists. Indeed, the Stockton pilot is being funded by the Economic Security Project, a San Francisco coalition of investors, activists and entrepreneurs, which was founded by Chris Hughes, Dorian Warren and Natalie Foster.

I first met Natalie many years ago, when I was setting up Impossible and she was running an organization trying to guide legislation around the sharing economy. A smart and impassioned social campaigner, Natalie became increasingly concerned by the shifting, insecure job landscape that the 'sharing' and 'gig' economies were creating. People are increasingly juggling multiple small part-time roles, without maintaining any underlying benefits or job protections.

Foster now advocates for 'portable benefits' alongside spearheading the UBI pilots. 'Cities are laboratories of democracy,' she says. The Stockton experiment has 'opened up the political imagination for what's possible', inspiring pilots in other cities across the US. 'To me this is the size of a solution that actually starts to match the size of the problem,' she tells me.

Still theoretical but like work benefits, just tied to an individual rather than the employer. They can be taken from job to job and, crucially, are universal, meaning they can be provided where an employer is otherwise unwilling or unable to do so.

Meanwhile, UBI is also being looked to as a more efficient way to address poverty than traditional aid schemes. Otjivero is a small village east of Namibia's capital where another UBI experiment ran for two years from 2008, with extremely positive results: household poverty was halved, malnutrition decreased by a quarter, income generating activities increased, overall crime nearly halved, and school attendance nearly doubled. Similarly, a UBI pilot by Broadway

Charity, seeking to address long-term homelessness in London, successfully housed nine out of thirteen of the people in the pilot, at a seventh of the normal cost to social services.

Part of UBI's surprisingly wide appeal to people across the political spectrum – from Dr King to the king of laissez-faire economics Milton Friedman – comes from the fact that it seeks to make welfare both kinder and more efficient. Reflecting on the Broadway pilot, *The Economist* concluded, 'The most efficient way to spend money on the homeless might be to give it to them.'

> The difficulty lies, not in the new ideas, but in escaping from the old ones.
> John Maynard Keynes, *The General Theory of Employment, Interest and Money*

In *Affluence Without Abundance*, anthropologist James Suzman argues that our work ethic is a largely modern phenomenon, shaped by farming, religion and politics. The San were very happy working a few hours a day, then using 'their free time to make music, create art, make jewelry, tell stories, play games, relax and socialize'.

UBI is not without critics. Perhaps because it conflicts with a long-ingrained sense of a 'work ethic', it is often not as popular as its advocates would hope. Keynes warned that we might find ourselves addicted to work: 'There is no country and no people, I think, who can look forward to the age of leisure and of abundance without a dread. For we have been trained too long to strive and not to enjoy,' he wrote in 1930. Although studies from Alaska and Finland suggest otherwise, people often fear that UBI will breed laziness, apathy and a refusal to work.

A brief mention of 'basic income' in AOC's Green New Deal opened the floodgates to conservative uproar. Brian Steensland, who has studied the history of basic income, compares the cultural resistance UBI meets in the US today to resistance to women's suffrage or equal rights for racial minorities in the past: i.e. not impossible to surpass, but going up against long-held and deeply established ideas.

When put to a vote in Switzerland – using their direct democracy system of regular referendums – the Bill was not passed. Some critics said the proposal was too expensive: a monthly income of 2,500 Swiss francs for adults and 625 SF for children, would cost the country a third of its GDP. My Swiss friend Osiris Roost said that cultural resistance played a part: 'Switzerland is a country where craftsmanship and work ethic are highly valued.'

That said, a 2016 poll found that the majority of people across the EU said they would definitely or probably vote for a universal basic income initiative and as the evidence and research from UBI pilots stacks up, and inequality and automation continue to grow, we may expect to see a shift in attitudes.

> You must not lose faith in humanity. Humanity is an ocean; if a few drops of the ocean are dirty, the ocean does not become dirty.
>
> Mahatma Gandhi

The very fact that we have founded national and international welfare and aid programmes demonstrates that people do care: that we do want to look after our communities – and indeed know that we will be looked after in return should we find ourselves in need. Our systems are far more complex but we do, in many ways, fly like birds in a V. How can we evolve these kindness machines and fix their failings?

'The one thing I urge us all to do, is take more risks,' says Natalie Foster, who maintains that it's only through pilots and data that we will discover if and how UBI can work, and what its limitations might be. When Michael Tubbs's mother found out about his UBI pilot, she asked him, 'Isn't this risky?' He laughed and said, 'The worst-case scenario is that 125 families in Stockton are better off than they were yesterday, and the best-case scenario is that we open up a political space for big ideas.'

> It is a fallacy that a life without poverty is a
> privilege you have to work for, rather than
> a right we all deserve.
> Rutger Bregman, *Utopia for Realists*

Arguably, the hardest obstacle in making a shift to UBI is a cultural one. Tubbs said his drive for the Stockton UBI pilot was 'to do something not just in policy, but also in the narrative and imagination – changing the narrative around who's deserving.'

For a long time, we have been taught to normalize and disconnect from the suffering of others: whether it's the underpaid garment workers sewing our clothes, the refugees seeking sanctuary on our shorelines, or the increasing number of homeless people lining our streets and building tent cities.

Our political institutions often ask us to harden our hearts to these realities. An underlying narrative supports this: that wealth is a product of hard work – that it is always earned – and that poverty must be a consequence of apathy. Yet we know this is not true. Many of the hardest-working people I know and have met would be considered poor.

UBI demands a paradigm shift in our collective thinking: away from the idea of the 'deserving' and 'undeserving', towards a desire for greater equality of opportunity. UBI also offers to better value existing unpaid work such as care work, domestic work and volunteering, and make space for more of it.

When I spoke to the two researchers behind the Stockton experiment – Dr Amy Castro Baker, and Dr Stacia Martin-West – they referenced the UBI experiments in 1970s Seattle and Denver (SIME/ DIME) in which women were found to work less. 'If you're working three jobs, you don't have much time to show up for your neighbours,' Dr Amy told me.

Like Dr King's concept of a 'beloved community', Michael Tubbs's driving vision also hopes to facilitate a more cohesive and inclusive society. 'In the present day, folks are working themselves literally to death like the robots that we say will replace them, and have more stress and anxiety than dignity. We're not designed just to work all day and run a rat race. We're designed to be in community, to volunteer, to vote, to raise our kids. And I think the more inputs and investments we can give people to do those things, the better off we are as a community.'

Involuntary simplicity

For the challenge of climate change, we can still avoid the worst, but that requires that the richest – as soon as possible – find a way to set their use of resources and energy that step by step can be shared by 10–11 billion people by the end of this century.

Hans Rosling, *Guardian*, 'Population growth and climate change explained', 2013

We once took our daughter to the birthplace of her grandmother, Mozambique. Amidst that beautiful landscape – breasts of mountains rising out of the land – I was struck by the irony of any contemporary Western claims for 'sustainability'. I passed miles of mud huts – small homes, handmade from red clay, with solar panels propped up against them.

A month after I left Mozambique, the region was struck by Cyclone Idai which killed more than 1,200 people and immediately flattened the fragile huts, leaving people stranded and largely dependent on foreign aid: the EU gave over €15 million, the UK £12 million, and many private donations poured in.

The incidence and scale of cyclones (and wildfires, flooding and droughts) are understood to be increasing because of climate change. The communities hardest hit are often the ones that have contributed far less to this cause, and have less in the way of infrastructure or resources to deal with its aftermath.

Whilst 'higher-income' nations help lead the way to cleaner and greener horizons, it is also hard to see how the West can be the model of sustainability when compared with the radical (involuntary) simplicity that billions of people live in right now.

Which illustrates the bind we are in.

Is sustainability about needing better products, or simply needing less?

Was our presence there positive, spending money in the local community, or did it do more harm than good by travelling?

Is economic growth a blessing that has created solar panels, and offers the ability to afford humanitarian aid and perhaps UBI, or is growth the beating heart of the problem?

Chapter Thirteen
Walking Together

Patriarchy, ownership and deep change

A woman walks down the road, wearing a light blue silk tulle dress and a soft smile. In her hands she holds a large red fake flower. Slowly, gently, deliberately, she swings the flower towards the window of a car and the glass splinters. She continues walking, her peace growing as she takes aim for the next car window, and the next. Multicoloured vehicles line the road like a paintbox, hers to casually fracture.

The film, *Ever is Over All* (1997) – which Beyoncé paid homage to in her music video 'Hold Up' – was made by Swiss artist Pipilotti Rist, after she proposed an image of an old lady for a newspaper commission and it was declined by the editor – he said it would negatively affect sales. The film reminds me of the suffragettes' technique of smashing windows as a form of direct action that would get them arrested without hurting anyone, under the ethos, 'better broken windows than broken promises'. But in Rist's video, a policewoman crosses the lady's path and they exchange smiles.

Hold up

If I thought Food was a controversial topic . . . try the other F word! Feminism. It's a bit like the word God: it means different things to different people. So

often, with many diverse people in my life, you mention the F word and people become very opinionated. I've done it myself.

Yet, can we stop feeding division? No one is responsible for the history they are born into, and no one can choose which body they land into.

BUT we all have a responsibility to try and understand the system we find ourselves in, and our place (and privilege) within it. The reality is, we are still emerging out of an <u>unbalanced system</u> that has been very oppressive to certain groups of people for thousands of years.

This shouldn't be a battle of the sexes. This is a battle of ideas. Let's collectively battle with old ideas of hierarchy and dominance, and move into a new paradigm that recognizes all humans as having equal value regardless of their gender, ethnicity, or

Author and social activist bell hooks coined her evocative term 'white supremacist capitalist patriarchy' as a shortcut 'jargonistic' method of summing up all the interlocking systems of domination that she encountered and felt needed to be addressed and examined simultaneously. She maintained that the patriarchy (believing in male dominance and superiority) was an ideology which both men and women can believe in and act on.

socio-economic position, and freedom to express themselves authentically.

Gender and Green

> At any rate, when a subject is highly controversial
> – and any question about sex is that – one cannot
> hope to tell the truth. One can only show
> how one came to hold whatever opinion one
> does hold.
> Virginia Woolf, *A Room of One's Own*

So, what is the relationship between gender and ecology? There is much to be teased out here, although it's intimidating territory to walk into: issues around gender are thorny, complex, and evolving.

I am not going to suggest that there is a 'right' way to do things – or that I know what it is – but present my explorations and thoughts until this moment, as the loosening of the grip of the patriarchy around the world gives me reason for optimism for society, and the environment.

Essentially, what I want to argue is that the way we have historically treated the 'feminine' – and people who have been oppressed through other constructed hierarchies, including ethnicity and class – is interconnected with the way we have treated the planet in a few thousand years of the prevailing systems.

Meanwhile, multiple studies show that women have been at the forefront of environmental movements worldwide, are more likely to enact environmental legislation and participate in 'environmentally friendly' behaviour, and that increased female autonomy helps to stabilize birth rates, and improve agricultural yields.

In multifaceted ways, the rise of the feminine can be seen as an essential key in our dawning environmental consciousness, and we can be optimistic that we are in a process of transformation, deeper and wider than we might realize, from which everyone can benefit.

Unfinished business

'System change' is sometimes chanted like a mantra.

But what does that mean? What does that mean to you? A change in economics, a shift in politics, a switch in energy grids?

Or is the *system* actually something much more pernicious, deeper and ideological – rooted in the invisible beliefs and underlying assumptions we might not even realize we maintain?

The underlying system of recent history is based on separating and ranking groups of people to justify power imbalances. The systematic oppression of women, marginal groups, different ethnicities and poorer groups of people, did not magically lift when slavery was abolished, women won the vote, homosexuality was decriminalized, some of the colonies were handed back or a minimum wage was established.

Of course, those were landmark changes, a loosening of the shackles of a fucked-up system over time, but we still have a way to go.

Throughout this chapter I will focus on the connections between gender and the environment, but I am aware of the various <u>intersections of prejudice</u>. Who would you rather have been: an upper-class woman, or a working-class man, in the sixteenth century? Queen or male slave? The lens of gender is useful, but does not exist in a silo.

People from different backgrounds experience the world in different ways, and types of prejudice overlap. It seems logical enough but it wasn't until the late 1980s that this term was coined by the lawyer and scholar Kimberlé Crenshaw arguing that aspects of social identity such as class, ethnicity, sexuality, gender and more should be considered in feminist discourse. Although over thirty years old, it has found a burst in popularity and acknowledgement in the last half-decade.

In the UK women still represent less than a third of MPs, a fifth of the House of Lords, and are under-represented in high-paying jobs. In business in the US, female-founded teams receive only 2.2 per cent of venture capital investment (compared with 76 per cent of all-male teams). In this book it hasn't escaped my attention that many of the entrepreneurs are white men.

Globally the gender imbalance is dire: In 2018 women represented 12 per cent of heads of state, 18 per cent of ministers, 24 per cent of parliamentarians. Every ten minutes an adolescent girl dies as a result of violence.

Brought up in a family of three women in London, and starting my professional career very young in the female-centric industry of fashion, I came to this conversation late. I never saw my gender as a barrier.

It was a male friend who campaigned me to pay serious attention to the wider and systemic issues that remain. As I started to look at the data and open my eyes to the reality for women around the world

– especially when their sex intersects with other forms of prejudice – I realized that my experience was atypical.

'Mother Earth'

> Greetings, Nature, mother of all creation!
> Pliny the Elder, *Natural History*

For thousands of years, cultures have associated our planet with the feminine: perhaps because of its life-giving properties and association with fertility. 'Nature' comes from the Latin word 'natura' – meaning birth. 'Gaia' was the Greek deity, and primal Mother Earth goddess; 'Dame Nature' appeared in medieval plays; in Hinduism the Ganges River is personified by Goddess Ganga, the mother, and the goddess Bhūmī-Devī is 'Mother Earth'. Then of course, there are the long-standing colloquial references we still use today: 'THE EARTH IS MY MOTHER, I MUST LOOK AFTER HER,' reads my daughter's sweater.

The Earth is not of course intrinsically feminine. It cannot be gendered, just as both masculine and feminine qualities reside in all of us (regardless of the gender we present).

Yet the cultural lens of 'Mother Earth' might help us to better understand the Earth's oppression, complicit with that of real mothers. If for thousands of years society was neither willing nor collectively able to honour, respect and appreciate women's place as equal – making mothers, sisters and daughters property, slaves, without legal, financial or political rights – it sort of makes (painful) sense that society felt entitled to abuse the Earth it considers 'Mother'.

There have been times when the image of Earth as a 'kindly mother' was used to protect it. Greek and Roman writers, like Pliny and Ovid, had long decried mining as a brutal act of injury to perform on 'Mother Earth' and her reproductive organs.

In the sixteenth century, Edmund Spenser (author of *The Faerie Queene*) described mining as 'a cursed hand the quiet wombe, Of his great Grandmother with steele to wound' and in the seventeenth century John Milton in *Paradise Lost* said, 'Men also . . . Ransack'd the Center, and with impious hands Rifl'd the bowels of their mother Earth.'

Yet eventually, by and large, cultural and religious ideas in the West that had long revered 'Mother Nature' and presented divinity in pantheist and often feminine ways, were supplanted by the idea of a singular monotheist and usually white male God, and over time exploiting 'Mother Nature' grew to be culturally accepted, even justified.

Thy desire shall be to thy husband, and he shall
rule over thee.
Genesis 3: 16, King James Bible

In the Scientific Revolution of sixteenth- to seventeenth-century Europe, a new way of thinking about nature attempted to subvert these ancient 'feminine' metaphors, to justify exploration and mining in almost sexual terms. Francis Bacon – the English philosopher and legal counsel to Queen Elizabeth I – described nature as a 'wanton harlot' and advocated taming her with scientific method: 'I am come in very truth leading to you nature with all her children to bind her to your service and make her your slave.'

As a driving force behind the colonization of North America, Bacon helped export this European sentiment to the rest of the world. Native Americans often resisted the Europeans' treatment of the Earth, with familiar, feminine metaphors. 'You ask me to plow the ground! Shall I take a knife and tear at my mother's breast? You ask me to dig for stones! Shall I dig under her skin for bones? Then when I die, I cannot enter her body to be born again,' said Smohalla of the Columbia Basin Tribe in the early nineteenth century.

These metaphors have not exited the cultural and political imagination today. When he was asked by international journalists about environmental protection, Brazilian President Bolsonaro, who has encouraged extractive, economic development of the Amazon rainforest, said, 'Brazil is a virgin that every foreign pervert wants.'

The marginalization of women and the
destruction of biodiversity go hand in hand
Vandana Shiva, *Women's Indigenous Knowledge
and Biodiversity Conservation*

Maria Mies and Vandana Shiva argue that the contributions to society of the natural environment and of women doing unpaid domestic and care work are rendered invisible under capitalism, because they produce neither profits nor capital.

These lines of inquiry – often termed 'eco-feminism' – can feel very abstract, yet provide a helpful framework for understanding how women's rights, environmental rights, and civil rights movements have grown up together: different types of oppression are interconnected, and the success of one movement empowers another.

Eco-feminism also captures how deep and underlying beliefs shape our behaviour. Thus, instead of understanding environmentalism as a matter of designing better technology, using less plastic or even writing policies, we are presented with a vision of environmentalism which requires a deeper cultural shift in our world view.

> The system of patriarchy is a historic construct;
> it has a beginning; it will have an end. Its time
> seems to have nearly run its course – it no longer
> serves the needs of men or women and in its
> inextricable linkage to militarism, hierarchy, and
> racism it threatens the very existence of life
> on earth.
> Gerda Lerner, *The Creation of Patriarchy*

I am optimistic about our moment in history and the trajectory of change we are part of, because around the world, over the past few centuries, different ethnicities, minority groups and women have been winning increasing legal rights and power, whilst men are gaining freedom to fulfil new roles in society. As we begin to collectively expand and evolve our concept of the feminine (in men as well as women), it seems no accident that we are beginning to honour 'Mother Earth' – even if we should stop calling it that.

The paradox of empowerment

A woman's freedom begins in her wallet.
Simone de Beauvoir, *The Second Sex*

When people discuss gender equality they usually talk about getting equal representation and pay for women, without questioning the underlying systems themselves. This logic is the ethos of the World Economic Forum's 'gender gap index' which tracks the distance between male and female power according to four criteria: politics, education, health and economics.

The 2018 index found that women had reached 68 per cent parity overall. Whereas the health gap has been successfully closed in some countries, and the education gap is estimated to close in fourteen years, the political gap isn't expected to close for 107 years and the economic gap for 202 years.

The power of economics to liberate women was the centre of Virginia Woolf's eloquent and groundbreaking argument in *A Room of One's Own*: that until women get economic emancipation they will not receive cultural and creative emancipation: 'A woman must have money and a room of her own if she is to write fiction.' This, she believed, was more important than the political vote.

This philosophy was the heart of Professor Muhammad Yunus's micro-finance projects in Bangladesh which empowered women through small community loans; it is what I saw empower women in polygamous communities in Ghana, when they were able to earn money from shea nut butter; and it has been the driving ethos for me, as a working woman, and founder of social businesses.

Meanwhile numerous studies have demonstrated the positive impacts that gender parity can have on the governance of companies. A report in 2015 by McKinsey estimated that if women achieved gender parity with men in work, global output would increase by more than a quarter, although whether that is aligned with environmental sustainability is a different question.

Whilst getting women into positions of leadership represents a massive and essential leap forward – and a Trojan horse for deeper changes to be made – this emphasis often stops short of remembering that the very systems in which we seek to empower women – companies, governments, financial markets – were predominantly designed by white men of privilege, during a particular history.

Might we pause to imagine what the world would look like if 'feminine' values (in men and women) had been equally active in determining our world – its rules, systems, culture and conventions – for the last few thousand years? We might have gotten Shakespeare's sister and . . .

Balance

Through the use of the dominator and partnership models of social organization . . . we can also begin to transcend the conventional polarities between right and left, capitalism and communism, religion and secularism, and even masculinism and feminism . . . all the modern, post-Enlightenment movements for social justice, be they religious or secular, as well as the more recent feminist, peace, and ecology movements, are part of an underlying thrust for the transformation of a dominator to a partnership system.

Riane Eisler, *The Chalice and the Blade*

Virginia Woolf's provocative contention in *A Room of One's Own*, that William Shakespeare had a genius sister called Judith who 'was as adventurous, as imaginative, as agog to see the world as he [William] was. But she was not sent to school.' Not to be confused with Shakespears Sister: a pop duo founded by a former member of Bananarama.

In *The Chalice and the Blade*, Riane Eisler places past societies on a spectrum from 'dominator' to 'partnership' cultures (arguing that partnership cultures were more peaceful). This is a non-dualistic way of seeing the world, which she hoped would move us away from divisive binaries – such as 'right and left', 'capitalism and communism' and 'even masculinism and feminism'. Eisler identified modern movements such as feminism and environmentalism as part of 'an underlying thrust for the transformation of a dominator to a partnership system'.

Eisler's work inspired bell hooks who expanded, in *Teaching Community*: 'Dominator culture has tried to keep us all afraid, to make us choose safety instead of risk, sameness instead of diversity. Moving through that fear, finding out what connects us, revelling in our differences; this is the process that brings us closer, that gives us a world of shared values, of meaningful community.'

In an alternative universe, where women had been equally powerful as men for the last few millennia, things would likely look very different. How exactly we cannot know, but perhaps the study of gender-balanced ('partnership'), cultures can give us an indication.

Neither matrilineal nor patrilineal, the nomadic communities in Southern Africa can be considered some of the most sexually egalitarian human societies the planet has ever known, and offer hope that balance is possible.

Patricia Draper studied two !Kung communities – one nomadic, and one settled into an agricultural village – to see how the process of settling impacted gender roles and relationships, and community life more generally.

'In the bush, village space is small, circular, open, and highly intimate. Everyone in the camp can see (and often hear) everyone else virtually all of the time.' This visibility made 'hoarding virtually impossible' and also ensured that !Kung values (opposed to 'physical fighting and anger, ranking of individuals in terms of status, material wealth, and competition') were regularly reinforced by the group.

Though men and women had different roles to play, they often switched to help one another, and equal gender roles were taught to children from a young age: 'As children grow up there are few experiences which set one sex apart from the other.'

Draper compared their lives with a different !Kung group who had settled into an agricultural village called Mahopa. In several subtle ways, Draper found that the move from a nomadic life to a settled agricultural one resulted in the loss of female autonomy and influence.

She also found that the children were given different gendered roles: the young boys were responsible for animal herding, which would expose them to other communities and languages outside their villages. This made them generally better at speaking the local language, Bantu, than the girls, and consequently more likely to play political roles, negotiating with local groups, in later life.

Adults too performed more rigid gendered roles: whereas in the bush, 'adults of both sexes seem surprisingly willing to do the work of the opposite sex', in Mahopa, 'women's work is seen as "unworthy" of men'.

This is interesting, as it consolidates the view that gender (the way we understand 'male' and 'female' roles and stereotypical behaviour) is largely a social construct. Alice Clark also argues that the dynamics of the Industrial Revolution and Victorian values transformed and redefined 'male' and 'female' roles in England. Whereas in the seventeenth century, women would participate in similar work to men, running household businesses – 'brewing, dairy-work, the care of poultry and pigs, the production of vegetables and fruit, spinning flax and wool, nursing and doctoring' – with industrialization, work became located outside of the family space, in factories: leading to the domestication of women, and 'husband as breadwinner'.

Draper argued that the movement from nomadic life to settled life, for the !Kung, similarly encouraged the greater domestication of women, as the richer material lives in Mahopa, and more elaborate mud huts, demanded more of the women's time to look after.

Whereas !Kung bush-people had been egalitarian, Draper observed that 'ranking of individuals in terms of prestige and differential wealth has begun in the settled villages'. Men became ranked according to their status, and the concept of ownership (of a house, a goat or a child) became commonly connected to them, rather than shared equally between men and women as it had been in the bush.

The greater privacy of the mud huts – spaced apart with fencing for the animals – gave rise to the hoarding of objects, and greater material inequalities. Whereas in the bush physical violence was rare – to be quickly intercepted by the wider community – with the privacy of huts, domestic violence was said to become a problem.

Draper's study suggests that the movement from a nomadic to a settled life had a transformative impact on social relations. Most human societies, including our own, made this transition from nomadic to settled lives during the Agricultural Revolution, roughly 12,000 years ago. Although it is speculative to draw a comparison between the !Kung and ourselves, it is interesting to ponder the questions her research provokes.

Are rigid ideas of male and female 'roles' socially constructed?

Did private property create inequality and gendered power imbalances?

Is ownership complicit with the patriarchy?

The anthropologist Claude Lévi-Strauss argued that the Agricultural Revolution made children more valuable, as help was needed to manage the land, and therefore childbearing women became valuable commodities for trade. He maintained that it was through the consequent 'exchange of women' – when tribes would exchange, buy and sell women for marriage rights – that private property was first established.

In *The Creation of Patriarchy*, Gerda Lerner argues that this practice of exchanging women for marriage made them the first slaves. This history saw the intersection of racism, sexism and classism, as it was usually the women of lower classes or persecuted races who were the first to be enslaved, showing how different forms of oppression compound one another.

To own or not to own

The loyalty of school children, indigenous
knowledge, drinking water, the human genome –
it's all for sale.

Lewis Hyde, *The Gift*

In Sweden, I met a scientist working for Ownershift – a think tank dedicated to democratizing ownership, beginning with female ownership in Sweden. Their data revealed that 15 per cent of land is owned by females, and 30 per cent of women own stocks or start businesses.

Even these disproportionate amounts of female ownership have only been made possible by the centuries-long fight for women to have any right to own their own property. Equal ownership is the next step, but it is often pursued without questioning whether property ownership itself comes from the same patriarchal world view that once considered women property.

Terra Nullius stems from the Roman law *Res Nullius* ('nobody's thing') which held that unclaimed things (wild animals, lost slaves, abandoned structures) could be seized as property.

This same paradox befalls indigenous communities around the world. When British colonialists arrived in Australia in the seventeenth century, they used the fact that the indigenous communities did not have a concept of ownership as excuse to occupy: the law *Terra Nullius* meaning 'nobody's land' is believed to have been applied, even though the indigenous communities had been living there for at least 70,000 to 120,000 years.

Now indigenous communities around the world have fought – and continue to fight – for their own land rights. This is essential to re-balance power within the given structure, but many indigenous communities continue to philosophically maintain that they do not own the land. Jeffrey Lee, from the Djok Clan (who you may recall resisted pressure to sell land for uranium mining), refused offers to lease the land saying, 'I don't own the land, the land owns me.'

My friends Nick Tree and Lauren Wilce work with the Bundjalung indigenous matriarchal tribe, in Western Australia, who operate according to three laws: share, care, and always tell the truth. 'In general, in the matrilineal the highest authority is the most senior woman or women in the tribe,' says Nick. 'Certainly, you notice in matrilineal how the people walking around issuing orders are the old women.'

'No one owned the land – everyone is custodians or caretakers of the land. The concept of a human owning land is completely alien to their way of thinking as everyone comes from the land, the land is the mother. You can't own pieces of your mother. In their culture you can only own something that's an object you've made yourself; such as a tool or painting,' Lauren tells me.

We might not be able – or want – to move away from the ownership patterns we have long established, but can we shift something in our mindsets, to remember that the earth ultimately owns us?

> Both male and female human beings are basically
> nurturers. If we reintroduce the gift paradigm
> into our interpretation of the world, we will find
> our 'gift giver within'.
> Genevieve Vaughan, *For-Giving*

Genevieve Vaughan makes the case that the gift economy is the natural state of being, as all humans enter the world through their relationship to their mother (<u>though men can also 'mother'</u>), which 'lays down the patterns of communication and mutuality that are carried out in the gift economy'.

Arguably aka an involved/modern dad. The gendered terminology here gets a little confusing, I know.

I was childless when I first read Vaughan's work but it rang deep and true notes to me. It became ever more tangible, when I reflected on my own intense and unconditional love for my daughter Wylde. I want to give, endlessly, and without expectation of return. What in any case would a young child have to give back, beyond love, kisses and cuddles?

Meanwhile Wylde didn't naturally understand exchange. She doesn't always share perfectly and she certainly welcomes presents; but she genuinely loves giving. When playing shop, she would always give the customer fake money, alongside the object. So if the toy cost £1, and you bought it, she would give you £1 and the toy. To my dismay (later) she started taking the money in exchange for the toy. She had, of course, 'learned'.

Walking together

Another world is not only possible, she is
on her way.
Arundhati Roy, *War Talk*

I first met Putanny Luiza, one of the first female shamans of the Yawanawa indigenous tribe, in her village deep in the Amazon in 2014, after a four-day trip to get there. With large white feathers crowning her head, Putanny danced, free and forceful, at the centre of her village, encircled by the forest, lit by the moon and a large fire.

Occasionally she would pull people into her orbit, holding hands, making contact, and then she would let go, and go back into her own internal rhythm. I was dancing nearby and felt like a self-aware teenager at a school party. When I thanked Putanny the next day for letting us stay, she said, 'Come any time – we are simply guardians of the land.'

In the past, only Yawanawa men could become shamans and work with their plant medicine aya huasca. 'The women wouldn't even be heard. When men would speak, women would have been kept away, just listening. Women's opinions weren't taken into consideration,' says Putanny. But when a lady from

A psychoactive brew ('plant medicine') made primarily from the *Banisteriopsis caapi* vine. Symptoms often include vomiting ('getting well'), diarrhoea and muscle spasms followed by intense, often epiphanic spiritual revelations. 'Whereas drugs make you happy for a short time, then sad for a long time, plant medicines make you sick for a short time, then happy for a long time,' someone once told me.

371

Rio de Janeiro visited and started drinking ayahuasca, it got them thinking.

In 2004 Putanny, and her sister Julia, received a 'calling' to do the plant medicine work. They followed a one-year shamanic initiation: living in isolation from the community and drinking ayahuasca on a diet of no sex, no sugar and no water. Putanny and Julia became the first female Yawanawa shamans, and now all Yawanawa women play central roles in their community's spiritual work.

'This calling came to bring a balance to my people – between the masculine and feminine. We women are powerful, are warriors,' says Putanny. 'We have the ability to generate life. That is unique to us,' says Putanny. Her husband Chief Nixiwaka also reflects that the recent participation and empowerment of women 'has been a fundamental spiritual nourishment for Yawanawa'.

> Indigenous women have the chance to bring
> new strategies to this moment of political crisis
> and deadly policies.
> Célia Xakriabá, indigenous activist,
> *Al Jazeera*, 'Women take lead in Brazil's
> indigenous fight'

The journey of Putanny and her sister represents a wider shift that is happening across South America, in which more indigenous men and women are taking positions of leadership. In 2019, a record 136 indigenous candidates ran in national elections which saw Joenia Wapichana become the first indigenous woman elected to Brazil's congress.

It is interesting that Putanny – a powerful female leader who has helped rewrite history for her community – tells me, 'I'm not a feminist, I don't like this word. I guess when a woman says she's a feminist, she wants to be more than man, she wants to be superior.'

Instead Putanny is seeking balance: 'I think that men and women have both their own individual value. Women are now heard and are treated with respect . . . not more than men, but side by side with men. We can't walk just the masculine or the feminine, we need to walk together.'

Fewer than 20 per cent of women in the UK consider themselves feminists even though 80 per cent believe in equal rights.

Unmanly?

When you look at the data on women and the environment, some surprising statistics emerge.

Because of existing vulnerabilities, women are the ones most likely to be hardest hit by the fallout from climate change and are up to fourteen times more likely to die during natural disasters. Male survivors outnumbered women three to one after the Thailand 2004 tsunami.

Meanwhile, women are estimated to represent 60 to 80 per cent of the membership of environmental organizations and multiple studies show that they are more likely to practise environmentally positive attitudes and behaviour patterns, such as lower energy consumption, eating less meat, and recycling more.

Women produce 60 to 80 per cent of food crops in low-income countries, yet have less access to land rights, education, financial credit or technology. According to Project Drawdown, empowering women with equal resources would increase farm yields by 20 to 30 per cent, which has the potential to feed 150 million people, reduce the pressure on land for food, and save over 2 gigatons of CO_2 by 2050.

A detailed analysis of different countries found that having more women in positions of political leadership made a country more likely to enact environmental legislation. The report concluded, 'Gender equality should positively influence state environmentalism.'

The data is clear – yet the reasons are not. Different theorists have speculated as to why this 'gender gap' might exist. Many believe that it is largely a consequence of the cultural ideas we have inherited from the patriarchy: that being 'feminine' means being more selfless, and that women have been socialized to be more 'other-orientated' – to look after children when disaster hits, do domestic work, and perhaps opt to be vegetarian.

One paper, based on seven studies, argued that there is a perceived association between 'green' consumer products and 'femininity' and so some men are, perhaps subconsciously, trying to assert their masculinity by avoiding green behaviour. In one test, men were found to be more likely to donate to an environmental charity with a masculine logo (blue and dark, with a howling wolf), than a feminine logo (green and light, with a tree).

> The idea that emasculated men try to reassert their masculinity through non-environmentally-friendly choices suggests that in addition to littering, wasting water, or using too much electricity, one could harm the environment merely by making men feel feminine.
>
> Aaron R. Brough and James E. B. Wilkie, *Journal of Consumer Research*, 'Is Eco-Friendly Unmanly? The Green-Feminine Stereo type and its Effect on Sustainable Consumption'

Is it cool to care about a warming world?

Beyond gender

I think we have to be very careful about how we interpret 'gender gaps'. Just as it might be tempting for some people to positively interpret the environmental behaviour gender gap to suggest that women are somehow fundamentally more proactive or 'altruistic'; so too can that type of gender stereotyping be used to justify other gender gaps – i.e. the argument that one of the reasons for the gender pay gap is because women are more 'agreeable'. We need to wonder why 'personality' differences exist.

'Biology doesn't render any one set of people superior to another – I think we need to be really cautious about biology and how we use it,' says black activist Marai Larasi, recalling how biological theories have been used in the past to justify scientific racism.

A study of more than 12 million participants found 'far more similarities than differences between genders', leading them to argue that gender roles are 'social constructs that vary significantly across time, context, and culture'.

If gender (unlike sex) is largely socially constructed, can we debunk it?

Is this happening?

The good news is that our understanding of gender has never been so fluid. As men and women push the boundaries of their expected roles, gender itself as a binary concept is being challenged. Facebook now has over seventy gender categories. This fluidity opens up the space for everyone to express themselves in new ways.

In the UK, three-quarters of people who take their own lives are men, and suicide is the biggest killer of young men. I'm not suggesting I know why this is the case, but whilst discussing gender equality for women, it's important not to lose sight of the issues that men also face in our current culture.

Did you know that clownfish, wrasse, mushroom corals, and even some chickens, can change sex? Which suggests sex is not some essential part of one's consciousness, but rather a physical biological reality. Can we get beyond gender?

The patriarchy limits us all. Men also suffer from the pressures of existing in rigid gender roles and stereotypes, and not being encouraged to explore the more 'feminine', emotional and/or paternal aspects of themselves; having to perform particular roles, dress in certain ways, and sometimes having fewer custody rights to their children.

As much as we need to open up roles for women, we also need to empower men to be equal fathers, 'feminists' and feminine if they want to be. We are all suffering from the delusions and trappings of the patriarchy. We will all benefit from moving into greater balance.

The anthropologist Peggy Reeves Sanday studied 112 societies to see how their creation myths impacted their gender attitudes, and found that in societies with a female deity, fathers were nearly four times as likely to look after infants, compared to those with a male deity.

In many countries the law is now evolving, in a positive direction in terms of generating equal paternal and maternal rights: in the UK parents can split maternity and paternity leave. Sweden has been encouraging fathers to share subsidized parental leave with mothers since 1971. (Incidentally the same year most women in Switzerland won the vote!)

When I first became pregnant with my daughter, my partner and I wrote and signed a contract on a Post-it Note, part in jest (but in spirit *very* seriously!). On it we agreed to strive to share all child responsibilities equally, fifty-fifty, between us, with the exception of breastfeeding which I generously agreed to do all of.

Sex and babies

She hasn't got balls – she's got a bleeding vagina,
she's about to chair the PTA committee meeting.
She's a fucking warrior.
Jonathan Pie, *My Problem with Mansplaining*

A lot of this boils down to sex and babies. Rather than being celebrated as the incubators of life, women have historically been disadvantaged by their biology: often physically weaker; bleeding every month (which makes work and school harder); then getting pregnant – often repeatedly – and taking responsibility for the children; sometimes dying in childbirth. If they are lucky enough to survive birthing multiple children, the hormonal roller coaster of the menopause awaits.

In eighteenth-century England, if a woman got pregnant out of wedlock (even if she had been raped) she would usually lose her job and her home, and was legally fully responsible for the child. Archives of these women's accounts show that they usually ended up begging on the streets or going to a workhouse where the infant mortality rate was 95 per cent.

Ironically, if the woman was married, and the couple separated, she would have no custody rights over her children. Marriage conveniently made the children (and the woman's finances) the property of men.

In the twenty-first century, alongside the push to give men and women equal rights and responsibilities for their children, we also have to consider the role that pregnancies play. Giving people the tools to choose how many children they have is an essential part of affording women empowerment around the world. It is also a move that will have an environmental side-effect, by stabilizing birth rates.

Population growth makes a regular appearance in environmental discourse and is a controversial topic. The suggestion that it needs to be curbed summons up ideas of state-controlled methods of birth control, disastrous political policies, and racial ideas about who has the right to have more children.

There are also places in the world where this conversation is irrelevant: some communities are actively trying to have *more* children to resist genocide – like my friends from the Yawanawa tribe in Brazil, whose population was decimated through colonialism. Population growth must also be considered second only to consumption growth and inequality: the average person in the USA has a carbon footprint equivalent to 150 people from Ethiopia.

There is a real danger that artificial fears around population growth can drive eco-fascism and environmental racism. 'What really scares me,' said Naomi Klein, 'is when a racist person stops denying climate change, their response is, we need to protect our own . . . it's going to make us nostalgic for climate change deniers.'

So, let's be clear, this isn't about saying who should be having children. And there isn't a 'population bomb' scare – we are already close to 'peak child' and world population is expected to level off before the end of the century.

A best-selling book written in 1967 by Anne and Paul Ehrlich, whose alarmist predictions of mass famines caused by over-population did not materialize.

Hans Rosling coined the phrase 'peak child' to describe the moment when we have the maximum number of children in the world (population will likely continue to grow for a while longer because of higher life expectancies, then level off).

379

But when and where it levels off remains unclear. The global population is still growing – with the UN estimating it to be somewhere between 9.8 and nearly 13 billion in 2100. There is a strong, but delicate and nuanced, case to be made that if you empower women through better education and family planning services, then not only do you help women, but as a side-effect you might also stabilize global population sooner and lower.

> Making a fortune and bearing thirteen children
> – no human being could stand it.
> Virginia Woolf, *A Room of One's Own*

There aren't many women who want to naturally birth thirteen children, and many (like the Angolan mother of my friend Ana) die in the process. When people gain access to contraception, and sex education, they usually choose to have fewer children: hence why the birth rate is actually falling in many nations.

Marai Larasi tells me that her Jamaican grandmother gave birth to thirteen children: 'She loved them all . . . and she still mourned for those who had passed away. But it had been hard on her and her body, and so she also said that if she had had access to birth control she probably would have had fewer children.'

According to the World Health Organization, 214 million women say they want to decide when they become pregnant but aren't using contraception, and in Africa 24 per cent of women of reproductive age have an unmet need for modern contraception.

If you combined the impact of 'family planning' and 'female education' in Project Drawdown's list of climate solutions, they would arrive at first place, because when women are educated and have access to contraception, they often choose to have children later, or have fewer. 'Close the gaps on access to education and family planning, and by mid-century we may have a billion fewer people inhabiting earth than we would if we do nothing more,' says lead author Katharine Wilkinson.

Meanwhile, a burgeoning number of environmentalists are also choosing not to have children, because they do not want to put further strain on the environment and/or because they are too worried about the uncertain future faced by children. When I asked the World Land Trust's John Burton what else humans can do to improve their environmental impact his response was stark: 'Don't reproduce yourself!'

I feel that the choice to have children is an essential and important part of being alive; for every person who wants to have four, there will be another who wants to have none.

In fact, I think choice is the central issue: empowering people so that they can choose.

Mind how you think

To step outside of patriarchal thought means . . .
being critical toward our own thought, which is,
after all, thought trained in the patriarchal tradition.
Gerda Lerner, *The Creation of Patriarchy*

If we focus only on our disadvantage, we may not even be aware of our privileges. I asked Marai how I could check myself for my white privilege. This was the advice she gave me, which I think provides a helpful analogy for all of us who carry some kind of advantage:

'I am a non-disabled woman and so, for example, I routinely travel through the streets of London navigating physical space without any consideration about how some disabled people have to think about space in a particular way. I am not always mindful of that privilege as a non-disabled woman, it doesn't mean that it is not there. The thing is, the responsibility for disrupting that privilege rests with me, not with disabled people.

'Privilege is privilege and exists whether people are feeling it or not. It is important for white activists, white thinkers etc., to start from a place of acknowledging that they have privilege, and asking, what do I do to disrupt that privilege and what do I do to use that privilege in the advancement of equality?'

Perhaps, you are convinced of the need to move beyond the patriarchy/'dominator culture'. What does that look like? Of course, there are some clear and tangible steps that can be taken, in terms of the organizations we support – and the laws we try to implement. Yet, one of the most important things we need to do, is to check our own thinking, and critically, the way we interact with the people in our lives.

Navigating patriarchy is exhausting . . . and some
women's survival strategies can end up including
behaviours that are (arguably) misogynistic.

Marai Larasi

This applies to all of us. As Gerda Lerner argued in
The Creation of Patriarchy, women played proactive and
complicit roles in facilitating a patriarchal system,
therefore women also need to examine themselves for
internalized misogyny: for the ways that we might disrespect other women; to make sure we are not placing
'patriarchal' values – such as dominance, competition
and rationalism – as superior to stereotypically 'feminine' values.

In 2017, actress Anne Hathaway was brave enough
to admit that she had recognized internalized misogyny in her own thinking. After working with
(female) director Lone Scherfig, Hathaway said, 'I am
to this day scared that the reason I didn't trust her the
way I trust some of the other directors I work with is
because she's a woman. When I get a script, when I
see a first film directed by a woman, I have in the past
focused on what was wrong with it,' she said, pledging
to consciously work against this bias.

Composting

Whatever sameness I've noted in my
relationships with women is not the sameness of
Woman, and certainly not the sameness of parts.
Rather, it is the shared, crushing understanding
of what it means to live in a patriarchy.
Maggie Nelson, *The Argonauts*

The patriarchy is so subtle, pervasive and long-standing, it can feel omnipresent yet invisible, hard to grasp. Consider the cultural normative ways we continue to make the masculine superior: women typically losing their surnames when they marry, male surnames being passed to children, all-male sports dominating the landscape, God so often appearing as a He. Meanwhile our culture continues to prize strength, competition and hierarchy. Let's try to break the windows in our minds.

It's understandable that some feminists get angry; I've found myself angry at moments, but that doesn't feel like the right way to escape this bind. As bell hooks recalled the Buddhist monk Thich Nhat Hanh advising her: 'Hold on to your anger, and use it as compost for your garden.'

Chapter Fourteen
Falling in Love (Again)

Nature, mental health and coming home

Six people walk around the forest in Taiwan, erotically touching the plants. The common ferns were once valued by local tribes but then neglected by Japanese colonists. The 'eco-queer' video, *Pteridophilia*, by Chinese artist Zheng Bo, is playing in a lecture hall in San Francisco, California.

Turtle Island

Californian dream, the land of paradox: world leader in progressive legislation, holy land for technology, the silver screen, Kundalini gurus and New Age thinking, the fifth-richest economy in the world. Hidden under the folds of its beautiful landscape and culture is a history less celebratory.

California's name is thought to be a reference to the mythical island in the 1510 story 'Queen Calafia' – a remote and exotic kingdom rich in pearls and gold, filled with 'beautiful black women'. An estimated 300,000 Native Americans had been living on the west coast of the land they called Turtle Island – now more often known as the United States – for at least 10,000 to 12,000 years, before James Marshall, from England, discovered a flint of gold in a river there in 1848, and word quickly spread. The subsequent Gold Rush,

The name used by many Native American and First Nation people, to describe the land now often known as North America – based on numerous creation stories. It is an idea found around the world, that the Earth or cosmos is sustained by a giant turtle.

and timber industry, decimated the local environment and facilitated a genocide such that within fifty years the indigenous population had been reduced by nearly 95 per cent.

Western society long assumed itself superior: the colonizers were 'discoverers' of a 'new world', as if indigenous people were just part of the scenery. They othered and 'civilized' indigenous people, exhibiting them in human zoos and International Exhibitions, as examples of primitivism. 'The pygmies are very low in the human scale,' read a *New York Times* article in 1904 when Ota Benga, a Mbuti pygmy man, was displayed in a cage with a monkey in the Bronx Zoo.

Yet less than two centuries since colonial expansion, Western society is discovering its own hubris, as our scientists tell us that we are on the path to making our only home uninhabitable, fuelling a rate of mass extinction not seen in millions of years. Now Western society is increasingly looking to indigenous traditions to offer a pathway to better land management, with no less than the World Bank arguing that 'the loss of [indigenous] cultural and spiritual identity and ancestral knowledge [is] as serious a threat as the massive extinction of species on Earth'.

There are 300 million indigenous people worldwide, from tropical rainforests to the Arctic Circle. Their 5,000 distinct cultures represent 95 per cent of the world's cultural diversity and hold treasure troves of ancient ancestral knowledge, about how to interact sustainably with nature, and different 'non-materialistic' world views.

Indigenous territories encompass over a quarter of the world's land, in areas that contain 80 per cent of its biodiversity. The World Bank and the United Nations have called for indigenous communities to be leading participants in conservation efforts. The UN's biodiversity report in 2019 found that whereas three quarters of land environment and 66 per cent of the marine environment had been significantly altered by humans, these trends were 'less severe or avoided in areas held or managed by Indigenous Peoples and Local Communities'.

In 2017, we met with the Yawanawa again, this time in the Algarve, Portugal. The chief, Nixiwaka, spoke of the particular resonance of being in Portugal: 'I feel quite emotional being here. Around 550 years ago we were living in the forest and in harmony with nature. We hadn't met anyone except other tribes. Then the boats arrived from Portugal. The population of indigenous people in South America was reduced from 8 million to 500,000.' After the later influx of missionaries, the Yawanawa population dropped to fewer than seventy people, before the community rebelled against the evangelizing and sought to reclaim their language, songs and culture. The Yawanawa are now in their thousands.

The Yawanawa are attempting to reconcile our disparate cultures. Whilst fighting to reclaim their traditions, they have extraordinary grace and forgiveness enough to acknowledge what they have gained from the West: 'Now I am wearing flip-flops, a watch and I speak Portuguese. We play our music using

Western instruments,' says Nixiwaka. They travel the world, able to share their knowledge, because of what Putanny jests is 'the big giant metal bird with people, food and a bathroom inside'.

'I don't want to dwell on the past. Our cultures have come together now and there is so much we can learn from each other,' insists Nixiwaka each time we meet. 'Climate change is clear. In our Amazon forest, there are so many big changes, in our rivers, our plants, our forests. The only thing that can fix it is awareness and consciousness. In the same way that human beings had enough knowledge to create this problem, we will have the necessary knowledge to get it back, to recover it.'

Deep time

A deep time awareness might help us see
ourselves as part of a web of gift, inheritance
and legacy stretching over millions of years
past and millions to come. When viewed in
deep time, things come alive, that seemed inert.
New responsibilities declare themselves . . . Ice
breathes. Rock has tides. Mountains ebb and
flow. Stone pulses. We live on a restless Earth.
Robert Macfarlane, *Underland*

My daughter's grandparents' home in the Algarve, Portugal, was burning. They had been up all night every night for a week, watching as the wildfires raged in the valley forests around them. Checking the movement of the wind, in case it came their way. Taking turns at night to walk around their garden, putting out any spark that might float by, soaking the land with the water they had saved and prepared for this moment.

My daughter's father arrived. Breaking the police cordons to walk up through the mountains to his family home. His parents are fierce but they were exhausted. Together, they managed to save the land.

Wylde's great-great-grandfather was an African slave: sold and sent to Brazil. In Brazil, he met an indigenous woman and they had a daughter. Their daughter had a son (Wylde's grandfather), Ademir, who travelled to Paris and met a Portuguese woman, Paula. Paula was descended from the Portuguese slave-ship builders who had sent people like Wylde's great-great-grandfather to Brazil. She's a spirited, moral woman, and she rebelled against the racism implicit in her family's history. Paula and Ademir had two sons, one of whom is Wylde's father.

So in Wylde's blood run all facets of this painful part of our history: the African slaves, the Portuguese colonizers, the indigenous of the land they conquered, not to mention my mix of Celtic, Neanderthal and Anglo-Saxon blood. If that brutal history hadn't happened, Wylde wouldn't exist. Neither perhaps would you or I, as we are all carrying complex stories and bloodlines. Looked at in deep time, humanity is one big family who have fallen out throughout history. We have the ability to reconcile.

The Yawanawa's attitude of reconciliation gives me immense hope. The spirit of the gift is not about giving an object or even giving time or skills. The spirit of the gift, in its most powerful form, is the spirit to for-give. To for-give, rather than for-get. To know your history, to know the pain of your ancestors, to learn from the past, but to be able to let go and move forward.

We were all indigenous once. For many societies, the relationship with nature that our ancestors had has all but gone. Through centuries of colonization and industrialization, we have destroyed our own indigenous cultures, and animistic traditions, intellectually divorcing ourselves from our own ultimate dependence on the natural world around us.

Can we go full circle?

We may not be able, or want, to return to gatherer-hunter lifestyles. I don't seek to romanticize the past, and indeed there are enormous and transformative benefits which 'development' has brought, but can we realign our mindsets, value systems and priorities with older ways of being?

Can we use indigenous values to reshape modern systems? What would 'share, care and always tell the truth' look like in modern politics? Can we write laws for ecosystems?

Whilst chasing the future, what can we learn from the past?

The many different indigenous communities around the world have diverse practices but a few interwoven themes come up repeatedly, many of which we have looked at in Part Four of this book: cultures of sharing, simplicity and intimate knowledge of the natural environment.

Many indigenous cultures also have a more expansive sense of their place in time: a veneration for their ancestors, and a commitment to future generations. The Iroquois, for example, teach that they should always consider the next seven generations to come: thinking 140 years into the future.

These values are often rooted in an appreciation of a sacred living universe, ever available to us, which is perhaps the most important mind-shift required in response to the environmental crisis. And it turns out that making this shift just might make us happier too.

Biophilia

The way we see the world shapes the way we treat it. If a mountain is a deity, not a pile of ore; if a river is one of the veins of the land, not potential irrigation water; if a forest is a sacred grove, not timber; if other species are our biological kin, not resources; or if the planet is our mother, not an opportunity – then we will treat each other with greater respect.
David Suzuki, *The David Suzuki Reader*

Stemming from the Greek *bios* (organic life) and *philia* (love), the term biophilia was coined by psychologist Eric Fromm to describe 'the connections that human beings subconsciously seek with the rest of life'. Also refers to living plants as wallpaper (more please); and an album by Bjork.

When I was sixteen I travelled, alone, to Japan for a month. It was a rite of passage in many ways: I wanted to explore the world, to experience a culture far removed from my own and be challenged by solitude. In Tokyo at night, I landed like an alien on another planet. I dragged my suitcase across the city, straight to the cinema to watch Sofia Coppola's *Lost in Translation*, then walked out as if into the film set.

At the end of that trip I travelled to the mountains in Yamagata in the north of Japan. Asked if I wanted to meet a Yamabushi 'mountain worshipper', I laughed at the idea. I was a city slicker. Growing up in London, I had always found nature pretty tedious. It made me think of spindly trees squabbling for pavement cracks; the uncomfortable feeling at sunrise in an overheated tent; endless trudging through mud to claim a Bronze Duke of Edinburgh prize I didn't even want; or the burning heartache of running through a park, painfully reminiscent of running for the bus. The most interesting nature I'd noticed was a bumpy old tree near my primary school which was swallowing the fence.

Yet the next day in Yamagata, immersed in the majesty of that extraordinary landscape, I found myself seeing the world in a whole new way. Everything came alive. Entranced by the patterns created by the water's moving surface, I could have watched the river run for hours. The trees, centuries old and as wide as houses: totems touching the sky.

Did you know that if your DNA was strung together in a single line, it would stretch 10 billion miles – reaching as far as Pluto?! (Thanks to Bill Bryson for that fact.)

Nature (that reified, mythical thing over yonder
in the mountains, in our <u>DNA</u>, wherever)
dissolves when we look directly at it.
Timothy Morton, 'Queer Ecology'

The Three Mountains of Dewa are the oldest place of mountain worship in Japan, where people have trekked for over 1,400 years. The Yamabushi follow an ancient ascetic religion which combines mountain worship with Buddhism, Shintoism and Taoism. They believe that communing with nature for long periods of time can lead to enlightenment.

After my day in that landscape, the Yamabushi mountain ascetics' philosophy made more sense than any of the ad-hoc Sunday school classes I had attended growing up. If we don't worship the natural world around us, in its awe-inspiring beauty, its ability to slowly sustain or quickly destroy life, what really is there to worship?

> Empirically speaking, we are made from star stuff. Why aren't we talking more about that?
> Maggie Nelson, *The Argonauts*

Cuban-American artist Ana Mendieta produced over two hundred works of art using the Earth and her body as her sculptural medium. These 'Earth-body' works see her silhouette inscribed with rocks, flowers, sticks or blood, to remove the distance between her self and what she called the 'womb' of nature. Ana once said about her artwork, 'I have thrown myself into the very elements that produced me.' All art can only ever aspire to be as powerful as nature itself.

From the Inuit in the north, to the shamanic communities of the Americas, the Maori in New Zealand, Jain and Hindu practices in Asia, the San of southern Africa or the pagans in Europe, most cultures have at one time held great reverence for the natural world.

This was often highly pragmatic: without the insulation provided by imports, modern medicines or refrigerators, indigenous and gatherer-hunter communities were intimately dependent on and therefore respectful of their environment. They carried encyclopedic knowledge – passed down inter-generationally – about the natural world around them: which plants were medicines, which poisons, how to hunt and manage the land to increase biodiversity.

Today, we can regain much of that practical knowledge, and arguably much more, with modern science. Yet the power of the indigenous relationship to the natural world went beyond the pragmatic: their respect for nature was rooted in an understanding that humanity was an integral part of it.

One touch of nature makes the whole world kin.
William Shakespeare, *Troilus and Cressida*

The concept of people being intrinsically part of, and co-dependent with, a complex and non-hierarchical web of all human and non-human life, appears again and again in different cultures and indigenous teachings. Mihilakawna Pomo elder Lucy Smith recalled her mother saying, 'The plants, animals, birds – everything on this Earth – they are our relatives and we'd better know how to act around them or they'll get after us.'

Falling in Love (Again)

This world view offers us a reality check.

Millions of bacteria in your stomach are right now helping you to digest; you are breathing the oxygen created by plants around you, digesting the fruits from a tree or the flesh from another animal's back, and drinking water that has travelled through the sky, to the inside of mountains, through a badger's intestinal tract, through a peasant farmer's bladder, through a flower, into seas, coursing round for millions of years, before it runs through you.

You come from the stars, you are grown through the elements, and when you die, all your cells will transmute to minerals and nutrients, which will feed other ecosystems and rise up again in trees or flowers, thorns or plants, to be eaten by another animal, like you.

Beyond humanism

We stand somewhere between
The mountain and the Ant.
Chief Oren Lyons, 1977, speech at
United Nations

We are made of stardust, but are we all that important?

Humanism – the philosophy that gives humans agency and value, whilst prioritizing science – is often praised for fuelling some of our greatest social advances: the idea that every single person is valuable and deserving of dignity irrespective of their social

We are made up of the elemental residue of super-novas and dying stardust.

status, ability, sex, race or age has helped lead us away from witch burnings, public executions, child labour and slavery, among countless other horrors.

Yet is humanism actually a blind spot? We may find ourselves living in the most peaceful time for the largest proportion of humanity, but this is arguably the most violent time to nature. As doctors and scientists tell us that climate change is the biggest global health threat of the twenty-first century, might we pause to wonder if humanism sowed the seed of our downfall in a world where humans are not in fact omnipotent? Can we move beyond humanism – to speciesism, or planetarianism – and is that already under way?

In the 1940s, author and environmentalist Aldo Leopold advocated the establishment of 'land ethics' towards the natural world. This, he argued, would arise as a natural extension of the voluntary ethics we had already extended towards the human community – our inclination to build schools, roads, hospitals, practise philanthropy, the fact that we had outlawed slavery and given rights to women. Through evolution and ecological necessity, the 'boundaries of the community' were expanding to include the whole 'biotic community: soils, waters, plants and animals'.

Leopold was wary of waiting for government or business to solve our environmental problems. Instead he insisted that it was up to individuals to begin to practise 'voluntary decency' towards the world around them, and any land they were responsible for. This land ethics would emerge in the 'thinking mind' of the community.

> There must be some force behind conservation:
> More universal than profit, less awkward
> than government, less ephemeral than sport,
> something that reaches into all times and places.
> I can see only one such force: a respect for land
> as an organism.
> Aldo Leopold, 'The Land Ethic' from
> *A Sand County Almanac*

This might not mean hugging a tree, but rather taking interest in the natural world around us. Wherever you live in the world, life can be found everywhere, making small and quotidian gestures.

A flower pushing through concrete, a dance of flies, the feeling of dewy grass or muddy earth under your feet. A fox sleeps in my mum's garden. A black deer eats the roses. Birds share songs that the dinosaurs once heard. A squirrel dances quickly up one branch, then parachutes down to another (normal). My daughter notices a small golden snake crawling across the earth (unusual). An eagle dives down at a roundabout (very unusual) An adder, with grey diamonds lacing its back, swims in the ocean, where you just were (crazy). England – the place where nature had once seemed so boring to me – now teems with life in my eyes.

An accidental recipe for happiness?

Environmentalists . . . have been clear about what
people should not do . . . We know what we
are against; now we must explain what we
are for.
George Monbiot, *Feral*

Environmentalism, in this understanding, is neither a question of pragmatic survival nor a kind of
benevolent duty towards the bees and the dolphins.
It's about finding community in the broadest sense:
from our friends and family to the strangers we
meet along life's journey, the people we will never
meet in person but whose lives we touch upon
through our choices, to the infinitely complex
ecosystem in which we all live. It is a sense of community that extends beyond the boundaries of time,
to people, creatures and plants in the future. Many
people think of environmentalism as a form of sacrifice, but what if it is actually an opportunity for a
better standard of life?

Live in the sunshine, swim the sea,
drink the wild air's salubrity.
Ralph Waldo Emerson, 'Merlin's Song'

Research shows that exposure to nature has numerous
positive benefits for mental health: improving mood
and self-esteem, decreasing rates of depression, stress
and anxiety. Many mental health organizations now

recommend spending time in nature as a remedy to stress or depression. Named 'ecotherapy', exercising in nature is prescribed as an alternative or complement to medication or even sometimes as mainstream psychotherapy.

Our brains have historically been wired by the rhythms in the natural world, which nature can help reconnect us to. Likewise, a study at Oxford University found that meditation and mindfulness can be as effective at treating depression as antidepressant drugs.

Put simply, nature can help us get out of our heads. And if you catch yourself (as I do) saying you are going to visit 'nature' – remember you are nature! As are all the materials and invisible microbes that surround you, and the trees that you likely wipe your arse with.

Environ-mental

We believe the planet is healing – this healing
is within us. In our actions. We are first healing
ourselves, so that we might heal the world.
Putanny Luiza

For me, environmentalism is deeply tied up with mental health. It's all very well for me to say community requires trust, acceptance and sharing: but why are those so hard to achieve? It's all very well to promote sharing and simplicity, and say, 'All we need is less' – as I have a few times in this book – but what lack in ourselves does our addiction to *more* actually imply? What is the root cause of excessive consumption and materialism? What exactly are we trying to fix with 'retail therapy'?

Without wanting to fall into a gooey self-help mantra, I know that I can only be a good citizen, mother, sister, daughter, friend or partner, if I can first find peace and happiness myself. When I am feeling stressed, wrung out and empty, it is very hard for me to be considerate about my environmental impact, and I am more likely to offer only my anger and sadness to the people around me. When I feel positive, loved and empowered, I have so much more capacity to be considerate and generous.

> As above so below, as within so without.
> Hermes, *Hermetic Corpus*

My friend Iris began volunteering for environmental and humanitarian organizations when she was fifteen and then spent a decade working with them: deep within what she calls the 'campaign industrial complex'. She was smart, busy, passionate and committed to the cause: a self-professed 'workaholic'.

I hadn't seen Iris for a few years when I ran into her at a yoga class, and discovered she was the teacher's assistant. I was interested to hear her journey and it resonated with me. Though she had dedicated years to working on policies and international agreements, she knew 'on some more fundamental level, [the climate crisis] is not just about the wrong policies – it's about some deep rupture and fundamental disconnect from the Earth and the resources around us'.

Meanwhile, she found her colleagues unwilling to discuss the inner dimension to the crisis: 'The organizations I was working for were full of people who were broken and upset, they didn't know how to do anything expect "*other*" and create bad guys and targets, and work through strategies which were missing where this really comes from'. Alongside feeling completely depleted and emotionally exhausted by her work, she started to wonder if it was actually even making the climate crisis worse: by deepening divisions where we need to come together.

Iris's words reminded me of something my cousin Emma had said when I discussed the road protests with her. In a way, she felt the wider environmental movement had failed – the crisis they had spent decades trying to warn the world about had arrived. Emma put this down, in large part, to their 'oppositional' tactics, which she believed fundamentally created a sense of separation: 'Most people must have thought, you're an environmentalist, and I'm not like you – therefore I must not be an environmentalist.' Instead, these days, Emma is 'more interested in finding the common ground, finding where we can share values, because we are all facing the same threats, regardless of beliefs'.

'Change is far more than the right paper on the right desk at the right time,' Iris said. 'It's more about the human condition, and our internal state.' Now she teaches Kundalini yoga, alongside working in

Kundalini is an energy source that is understood to lie dormant at the base of the human spine, which this yoga seeks to awaken. It was apparently long mythologized that if Kundalini yoga was taught to more than one person at a time, the teacher would die within a year. Yogi Bhajan decided to test his luck and started teaching group classes in 1968 and he lived decades long enough to see the yoga explode in popularity around the world. I love it.

sustainability, trying to ask environmentalists, 'What is the work that brings together our inner and outer worlds?'

Iris pointed me towards the Collective Psychology Project set up by her friend Alex, which sees inner and outer crises as connected, and seeks to bring collective psychology into political and environmental action. The project, which is in its infancy, focuses on community, self-awareness, agency and belonging.

The heart is the great leveller. No matter our background, each of us has to grapple with invisible patterns and yearnings. We all have our stories – from childhood, from our ancestors – and they are not always easy to carry. Medical research has found that generational trauma is even carried in people's genes.

Yet each of us has the opportunity to take responsibility for our own healing, however that might look. We cannot always control our outer world, but we can seek control of our inner.

'In a sense, all responses to the current ecological climate are mad, or at least maddening,' writes the psychoanalyst Anouchka Grose. 'Take the threat seriously and risk succumbing to <u>sostalgia</u>, or blot it out and be accused of opting out of reality.'

A neologism that sums up the devastating effects of finding unease where you used to look for relief. A form of mental or existential distress brought about by climate change.

This is perhaps all the more pertinent as climate chaos is escalating the rates of grief, anxiety, fear and even suicide: creating a vicious cycle of mental and planetary ill-health. Anouchka tells me that the number of people she sees expressing mental health issues because of climate change has soared in recent years.

Can we escape the cycle?

I could, perhaps should, write a stand-up comedy about the number of 'self-help' things I have tried: you name it – I've probably had a go. At times, my mental health is a real struggle, especially when compounded by facing our environmental reality.

I certainly haven't got it all figured out yet, and probably never will. It's a process, not a destination.

Yet I feel blessed to be living in a time where there is unprecedented opportunity for healing: from Eastern traditions, indigenous African practices, Western psychotherapy, not to mention all the New Age approaches, wellness apps and increasing neuroscience advocating plant medicines.

There is no longer any real cultural taboo about whether you choose a <u>Freudian</u> therapist's couch or a simple daily meditation practice, yoga class, book, YouTube video or even a laugh with your friends. It is never selfish or weak to prioritize healing. It's potentially the most responsible, brave and generous thing you can do

'From error to error one discovers the entire truth.' Sigmund Freud was the founder of modern psychoanalysis. He said, 'One day, in retrospect, the years of struggle will strike you as the most beautiful.'

You are Nature

A few years after John Francis (who we met in Chapter Six) vowed in 1971 to only walk following the San Francisco oil spill, he got so tired of having to explain this that he decided to stop talking. His vow of silence lasted seventeen years, and he began to really listen.

Crossing America for seven years on foot, he says, 'I had the opportunity to visit with everybody, including people that my parents might have warned me against being with. Not speaking allowed me to learn to really listen to what someone has to say without saying, Wait a minute, I disagree!'

John broke his silence in 1994 at a conference to celebrate Earth Day, to share a simple message he had come – in those seventeen years – to realize. 'We are environment,' he told me emphatically. 'How we treat each other is really our first opportunity to treat the environment in a sustainable way.'

After doing a PhD in land resources, and listening to thousands of people on the road, John realized, 'We could write a regulation to fix problems, like pollution, loss of species, or climate change, or develop technology that might alleviate them and those are really good things. We shouldn't stop doing those things. But if we don't tackle the underlying cause, which is how we treat each other, the solutions will be a Band-Aid.'

Planting flowers

Die when I may, I want it said of me by those
who know me best, that I always plucked a
thistle and planted a flower when I thought a
flower would grow.

Abraham Lincoln

My family moved to the countryside and we started planting: herbs, tomatoes, flowers, strawberries, pumpkins, lettuce, flowers. Every weekend my fingernails would be blackened by the soil and at night as I lay in bed my mind raced back into the earth. Wylde loved it.

The abandoned greenhouse we were working in was full of weeds – 'pioneer plants' as they are more fondly known. Pioneer plants can teach you a lot about soil. Often overlooked, as 'weeds', they actually offer clues to the soil that we can learn from. Some are very valuable, and they all arguably deserve their place in the whole.

Needless to say, I tried to remove some of the thistles. I would anxiously keep trying to pull them out. Then every few days, ten or twenty more pushed their heads out again. It didn't matter how often I cut their heads off, they just came back, thicker and stronger. It was like playing Whack-A-Mole.

But if you run your fingers along the side of the weed you can reach down deep into the soil and search for the knobs and knuckles where they begin. It is only by going that deep, by pulling the root out, that it stops reappearing.

Weeding seems to me to be a good metaphor for life: unless we really get to the deeper knuckles and knob roots – the systemic causes – of our challenges, we will be forever weeding.

In my search for root causes to social and environmental issues I meandered through economics, politics, and technology, looking at different cultures and value systems across centuries and countries.

Through my twenties, my relentless utopian drive absorbed most of my money and time, and I ran a series of entrepreneurial experiments, many of which appeared to fail. I was obsessively driven by this belief that our collective lives can be better, and that there are pragmatic and small ways that we can each try to improve them. I saw business and technology as powerful tools to try and achieve change.

Yet – exhausted – I finally realized that the only thing I could seek to try and change was myself.

I'm still working on that.

This book seeks to offer more questions than it answers, because – at this critical moment in time – I believe the most important thing for us to do, is to listen to one another.

The exercise of writing this book has not alleviated my fear. Delving into the research and data, alongside increasingly crazy weather, at times I've felt more pessimistic than ever (which is ironic when fighting for optimism).

But in the process of exploring and writing this book, something extraordinary happened. Serendipitously.

A wide movement of people has emerged and banded together. The media has started reporting seriously on the issues. Companies have made bold commitments. Political action is emerging. Ideas that once seemed utopian are becoming mainstream.

Put simply, I do not feel part of a small niche.

Falling in Love (Again)

A lot of people care.

When armed with the information, I would say most, if not all, people care. Because not to care is to accept a premature death, and life is so interesting.

Ana Mendieta,
Imagen de Yagul,
1973.

Epilogue

Possibilities

The greatest threat to our planet is the belief that
someone else will save it.
Robert Swan, 2016, Social Innovation
Summit

This book has been dedicated, in large part, to the
belief in the power of the individual. Each of us has a
role to play in shaping our response to crisis, and each
of us carries the vibrant pulse of possibility. Yet, there
is of course only so much any individual can achieve;
solace may be found in recognizing our own limitations and ultimately letting go, ceding control to the
wider play of what some may call destiny.

According to biologist Lynn Margulis, it is the
fate of every successful species (such as humanity), to
'wipe itself out'. All species have natural limits, she
argues, and overpopulation will inevitably be curbed.
Charles Mann in *The Wizard and the Prophet* summa-
rizes: 'By luck or superior adaptation, a few species
manage to escape their limits, at least for a while . . .
Their populations grow at a terrific rate; they take
over large areas, engulfing their environment as
if no force opposed them. Then they hit a barrier.
They drown in their own wastes. They starve from
lack of food. Something figures out how to eat them.

Neither conservation nor technology has anything to do with biological reality. *Homo sapiens*, in Margulis's eyes, was just another briefly successful species.'

Fighting to stop climate change sees the ultimate survival instinct kick in. We are really fighting, selfishly perhaps, to save humanity and numerous other animal species dependent on us and on which we also depend. We are fighting change and its inevitability. Planet Earth will continue regardless. Perhaps, like the New Age billionaires investing in the quest for immortality, we are simply fighting death itself.

In my bleakest moments, I find it strangely reassuring to know that all great mass extinctions in the past have been followed by intense evolutionary moments: the space left by loss invites growth and gain in other ways. After the extinction of the dinosaurs (in all likelihood due to the impact of an asteroid) came the Age of Mammals, and so if the dinosaurs still roamed the planet, we would likely not be here – in this form as *Homo sapiens* – nor would the elephants, dolphins or monkeys. If we do precipitate our own mass extinction, new life forms will inevitably emerge.

I also find it reassuring to know that NASA have found water on other planets and therefore suspect the existence of life there. It is highly likely that there is other life somewhere else in this beautiful, vast universe.

Perhaps, whatever happens, it's really not the end of the world.

The future is dark, which is, on the whole, the
best thing the future can be, I think.
Virginia Woolf, journal entry, 1915

Yet is this fatalism the only reading of science? Have we figured it all out yet? When we look back at history we often patronize earlier cultures for their limited understanding of reality: believing the Earth was flat, or at the centre of the universe. There is so much we still do not understand: so much more to discover possible, as quantum physics elusively points to every day. We still don't understand consciousness itself.

Believing in limitless possibility is the ultimate optimism: it is the quest for Utopia that will take us further than our wildest dreams. What I have presented in this book are entirely realistic visions of a positive future within our grasp, using existing technology and ideas. Nothing I have suggested is impossible. Yet beyond what I have explored here, so much more can happen. The future will likely be composed of ideas that have not even been dreamt up yet.

Possibilities

> I have noticed that even people who claim that
> everything is predestined and that we can do
> nothing to change it look before they cross
> the road.
>
> Stephen Hawking, *Black Holes and Baby*
> *Universes and Other Essays*, 'Is Everything
> Determined?'

The day I met Stephen Hawking I was nervous. In his room there were photos of him flying in a spaceship, lifted out of his wheelchair through zero gravity, a wide smile on his face. Next to them was a photo of his cameo in *The Simpsons*. I had spent the previous few days cramming my mind with his work and writing up my little list of questions.

I asked the first – I think it was to do with quantum physics. It took him about ten minutes to type out a reply using one corner of his eyebrow, and I could see the answer appearing before it was finished. I knew how it was going to end, and I wanted to interrupt, to save him the effort of typing it all out – but that felt rude.

'You can read about that in my new book,' it said.

Here I was, wasting this man's time. Why hadn't I read all of his new book in time? The questions on my page shrank alongside my confidence. I asked him about free will and determinism.

'I think free will is an effective theory,' came his reply.

He then asked for a cup of tea. Flailing, I put my last question. It was the most important one.

'Do you believe anything is possible?'

I was asking the scientist who had studied the universe from molecules to black holes, who considered science his god, and advocated space travel, if he believed that anything is possible.

I was asking the man who lived twenty-seven times longer than the doctors predicted, if he believed that anything is possible.

Stephen gave a half-smile.

His son Tim gently laughed.

'That's a yes.'

Acknowledgements

I've always glanced at acknowledgements in books, and thought, wow, that's a long list of people. Now I understand why: it takes a village to write one. There are so many people I am grateful to for being part of the journey. I am sure I will accidentally miss some, so I'm very sorry if you were part of my village, but don't find your name here:

Mum, Elvie and K: words can't capture my appreciation for your love, thoughts, generosity and endless support. You are the best family one could wish for.

Ed Victor, who is no longer with us, but set me on this book path, thank you for your sparkly encouragement. Your words have followed me throughout. And Karolina who I feel so happy to have found a new home with.

Helen, thank you for all your dedication, advice and enthusiasm shaping this . . . you were invaluable when I needed it most. Harland, thank you for your beautiful artwork and witty words. You made a little dream come true for me.

Emily, Venetia, Natalie, Mary, John: what a team! Your work, guidance and patience have been essential, thank you. And Joanna whose idea this was to begin with.

James and Graham, you have influenced my thinking more than you probably know. Amia, Marai, Izabella, Emma and George, thank you for reading and offering insightful feedback. Emily and Eric, thank you for patiently answering all my endless climate science questions. Eliot, cheers for your sprinkling of humour.

Thank you so much to everyone who generously shared their thoughts and experiences with me and the people who helped facilitate these interviews. You've made this book whatever it is. I really appreciate that you took the time, and I hope I have reflected your work and views accurately. I enjoyed speaking with all of you, and learned many things along the way.

To the environmentalists who have dedicated their lives to our planet, and long guided me: Steve and EJF, John and WLT, John and Greenpeace. Everyone who has worked with me on the Impossible journey, including Scott at Pivotal Labs and Matthew at Freuds, your gifts will never be forgotten: Kate, Brian, Jimmy, Yunus, Ross, Nicola, Richard, Gabriel, Paula, Edward, Frank, Chelsea, Nicholas, Andrew, Helen, Mark, Dan, David, Simon and Mark at HSF, Jez, Genevieve, Hazel, Saskia, Perry, Joss, Peter, Tea! And all our Impossible community: for proving that kindness prevails.

Finally, to all the people who give more than a damn, for whom this book is written. To the people trying to dream and act in a better future. Thank you for caring.

Appendix

There have been at least five known mass extinctions in the past, of which the most dramatic was known as the Great Dying. Some 252 million years ago (9 to 19 million years before the dinosaurs appeared) a series of Siberian volcanic eruptions helped make the Earth 10 degrees Celsius (18° F) warmer. As temperatures rose, 95 per cent of marine life, and 70 per cent of land life, became extinct. We are currently adding carbon dioxide to the atmosphere at ten times the rate that happened during the Great Dying.

~~~

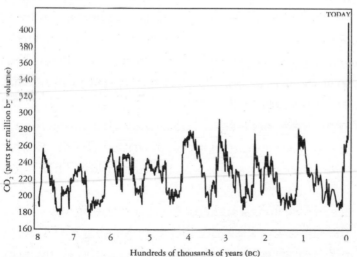

Carbon dioxide levels and global temperatures historically track together. The graph on the previous page shows the changes to CO_2 levels, caused primarily by deforestation and burning fossil fuels, since the Industrial Revolution.

In just a few decades we have managed to
disrupt the atmosphere as much as is achieved
by a change in orbit of the Earth that takes
100,000 years.
Emily Shuckburgh, Cambridge University
Climate Scientist

~~~

We also have a biodiversity crisis (driven by global warming, loss of habitat, hunting, over-fishing, pollution). Since 1970, wildlife populations have decreased on average by 60 per cent. A UN report, published in 2019, says the current rate of global species extinction is 'tens to hundreds' of times higher than the average over the last 10 million years, and is accelerating. One million species are threatened with extinction in the coming decades.

~~~

Ice in Greenland and Antarctica is melting at six times the rate it was a few decades ago. Melting ice from the polar caps causes *feedback loops*: for example, the more sea ice we lose, the less of the sun's heat is reflected back and the more heat is absorbed by a dark ocean – the warmer the oceans, the more ice we lose . . .

Scientists are unsure what the tipping point temperature will be to trigger uncontrollable 'runaway climate change' but some worry it may be less than the 2-degree warming threshold that was agreed in Paris 2015, and the IPCC now recommend we try to stay within 1.5 degrees of warming. The Great Dying is the poster child for runaway climate change.

~~

The final interview I did for this book, with someone leading a nuclear fusion company, brought me a surprise twist in the narrative. He was a climate sceptic. I tried to understand and interrogate his perspective, to be open-minded, as I have with the many other people I have met on this journey. It was soothing to consider. For a day or two, I was high on hope.

I did my due diligence, with the support and patient guidance of two Cambridge climate scientists (Emily and Eric) and landed back where I started. Unfortunately, the science is clear. Out of 69,406 authors of peer-reviewed scientific articles on global warming, only four rejected the hypothesis of human-induced climate change: meaning that 99.99 per cent agreed with it.

The basic facts are that CO_2 definitely absorbs heat energy that is leaving the Earth (known for well over a century) and since CO_2 has been increasing, climate is certain to warm. Sensible sceptics might argue over how much and over whether we should care, but not about whether it's happening at all.

If anyone argues that CO_2 doesn't cause warming then they are effectively denying large swathes of quantum theory and they also shouldn't 'believe' that the GPS system in their car works (I put '' around 'believe' because it drives us mad that people think science is about belief rather than evidence, but that's just me ranting).
Eric Wolff (Royal Society Professor of Earth Sciences, and Science Leader for the European Project for Ice Coring in Antarctica)

In a world of fake news, many people do not know who to trust, and want to believe it isn't as bad as the scientists say – it is hard to accept information so painful. A website called https://skepticalscience.com is good source for fact checking.

Of course, there remains a tiny, tiny chance that 99.99 per cent of scientists are wrong. But when it comes to risking life on Earth as we know it, does anyone really want to hedge their bets on hope?

Text Permissions

The publisher is grateful to the following for their kind permission to use the extracts in this book:

On p. 130: extract from 'Let Them Drown: The Violence of Othering in a Warming World', Naomi Klein, *London Review of Books*, Vol. 38, No. 11, June 2016. Adapted from the 2016 Edward W. Said London Lecture delivered by Naomi Klein © 2016; on p. 146: extract from 'Let Them Drown: The Violence of Othering in a Warming World', Naomi Klein, details as above; on p. 178: extract taken from an interview between Lily Cole and Mary Conrad; on p. 219: extract taken from interview by Christina Patterson with Arundhati Roy for the *Independent*, 'Arundhati Roy: What we need is a feral howl', The Big Interview, July 2017, © Arundhati Roy, from *Listening to Grass Hoppers*; on p. 242: extract taken from 'The destruction of the Earth is a crime. It should be prosecuted', by George Monbiot, March 2019 for the *Guardian* (Environment), https://www.theguardian.com/commentisfree/2019/mar/28/destruction-earth-crime-polly-higgins-ecocide-george-monbiot; on p. 290: extract taken from *Sapiens* by Yuval Noah Harari published by Vintage. Reproduced by permission of the Random House Group Ltd © 2015; on p. 388: 59 words from *Underland* by Robert Macfarlane (Penguin Books, 2019). Copyright © Robert Macfarlane, 2019.

Picture Permissions

The publisher is grateful to the following for their kind permission to use the images in this book:

On p. 1, courtesy of Olafur Eliasson and Minik Rosing, *Ice Watch*, 2014; on p. 15, courtesy of Olafur Eliasson, *Bankside*, outside Tate Modern, London, 2018; on p. 27, courtesy of Greg Lotus; on p. 28, © Gabriel Orozco, *Cats and Watermelons (Gatos y Sandias)*, 1992, Chromogenic colour print, 21 1/2 x 27 3/4 in./(54.6 x 70.5 cm), courtesy of the artist and Marian Goodman Gallery; on p. 36, © PA Images; on p. 45, *Haec-Vir/Hic-Mulier*, from *The Man-Woman*, 1620; on p. 90, exhibition view of 'The Black Image Corporation', a project by Theaster Gates, 20 September 2018–14 January 2019, Fondazione Prada Osservatorio, © Photo Delfino Sisto Legnani and Marco Cappelletti, courtesy of Fondazione Prada; on p. 211, Mary Conrad Sculpture, exhibition view of 'What is Shakespeare in a time of Climate Change', a project by Marfeco, 17 August–7 September 2019, Great Highway Gallery, San Francisco, photo by Lily Cole, courtesy of Mary Fernando Conrad; on p. 225, *Road Protest* © Andrew Testa/Panos Pictures; on p. 266, graph redrawn from 'Politics in 2014: The Coalition Dished Out Slogans, And Its Sentence Is Clear', by Lenore Taylor, *Guardian* Australia: www.theguardian.com/australia-news/2014/dec/19/politics-in-2014-the-coalition-dished-out-slogans-and-its-sentence-is-clear?CMP=share_btn_tw. Graph data originally from Australian Energy Market Operator; on p. 285, *U.N.*, 1995, Bodys Isek Kingelez, installation view of the exhibition 'Bodys Isek Digitale (1) Kingelez: Cit Dreams. MoMA, NY, 26 May

2018–1 January 2019. Digital photograph. Photographer: Denis Doorly (copyright: The Museum of Modern Art, New York). Object Number: IN2399.11. New York, Museum of Modern Art (MoMA). © 2020. Digital image, The Museum of Modern Art, New York/Scala, Florence; on p. 353, Pipilotti Rist, *Ever is Over All*, 1997, audio video installation (video still) © Pipilotti Rist. Courtesy the artist, Hauser & Wirth and Luhring Augustine; on p. 407 © ARS, NY and DACS, London 2019, Ana Mendieta, *Imagen de Yagul*, from the series *Silueta Works in Mexico 1973–1977*, 1973, chromogenic print; 20 x 13 3/8 in. (50.8 x 33.97 cm), San Francisco Museum of Modern Art, Purchase through a gift of Nancy and Steven Oliver, © 2020 The Estate of Ana Mendieta Collection, LLC. Courtesy Galerie Lelong & Co. / Licensed by Artists Rights Society (ARS), New York, photograph: Don Ross; on p. 415, graph redrawn from Composite CO_2 Record, based on data from Monnin et al. 2001 for 0–22 kyr BP, Petit et al. 1999; Pepin et al. 2001; Raynaud et al. 2005 for 22–393 kyr BP, Siegenthaler et al. 2005 for 393–416 kyr BP and 416–664 kyr BP, Luethi et al. 2008 for 664–800 kyr BP.

Further Reading

More suggestions can be found, slipped into the folds of the marginal notes. Endnotes can be found at: penguin.co.uk/content/dam/prh/penguin/Who-Cares-Wins-Notes.pdf

OMG

'Global Warming of 1.5°C,' United Nations IPCC report (2018): https://www.ipcc.ch/sr15/

P. Hawken, K. Wilkinson, J. Ravenhill, *Drawdown: The Most Comprehensive Plan Ever Proposed to Reverse Global Warming* (2008), Penguin, and free online version: https://www.drawdown.org/

C. Mann, *The Wizard and the Prophet* (2018), Picador

E. Shuckburgh et al., *Climate Change (A Ladybird Expert Book)* (2017), Penguin

Skeptical Science: https://skepticalscience.com/

D. Wallace-Wells, *The Uninhabitable Earth* (2019), Penguin

Part One – Chapter One

J. Suzman, *Affluence without Abundance* (2019), Bloomsbury

E. Ehrman, *Fashioned from Nature* (2018), Victoria & Albert Museum

Chapter Two

R. Carson, *Silent Spring* (2000), Penguin Modern Classics

J. Safran Foer, *We Are the Weather* (2019), Hamish Hamilton

I. Tree, *Wilding* (2018), Picador

D. Montgomery, *Growing a Revolution* (2017), W. W. Norton & Company

Chapter Three

M. Braungart and W. McDonough, *Cradle to Cradle* (2009), Vintage Digital

Part Two – Chapter Four

J. Ausubel, 'The Return of Nature: How Technology Liberates the Environment' *Breakthrough*, no.5 (summer 2015): https://thebreakthrough.org/journal/issue-5/the-return-of nature (cited by A. McAfee, *More from Less*)

T. Berners-Lee, *Web We Want*: https://webfoundation.org/2019/03/web-birthday-30/

D. Bradfield, *The Trust Manifesto* (2019), Portfolio Penguin

Chapter Five

N. Klein, *This Changes Everything* (2014), Penguin, and 'Let Them Drown', *London Review of Books* vol.38 no.11 (June 2016): https://www.lrb.co.uk/v38/n11/naomi-klein/let them-drown

S. Pinker, *Enlightenment Now* (2019), Penguin

Chapter Six

J. Francis, *Planetwalker: 22 Years of Walking, 17 Years of Silence* (2008), National Geographic

R. Garan, *The Orbital Perspective* (2015), Berrett-Koehler Publishers

E. Musk, 'HyperLoop Alpha' (white paper), (2013):

https://www.tesla.com/sites/default/files/blog_attachments/
hyperloop_alpha3.pdf
S. Hawking, *Brief Answers to the Big Questions* (2018), John Murray
P. Kingsley, *The New Odyssey: The Story of Europe's Refugee Crisis* (2016),
Guardian Faber Publishing

Chapter Seven

D. Keith, *A Case for Climate-Engineering* (2013), MIT Press
H. Thoreau, *Walden, Civil Disobedience and Other Writings* (2008), W. W.
Norton & Company
P. Kingsnorth, *Confessions of a Recovering Environmentalist* (2017), Faber
& Faber
G. Monbiot, *Feral* (2013), Penguin
M. Boyle, *The Way Home* (2019), Oneworld Publications

Part Three – Chapter Eight

G. Thunberg, *No One Is Too Small to Make a Difference* (2019), Penguin
E. Giffard et al., *Dark Mountain*, issue 13: https://dark-mountain.net/
shop/

Chapter Nine

K. Forrester and S. Smith, *Nature, Action and the Future* (2018), Cambridge University Press
M. Yunus and K. Weber, *Building Social Business* (2011), PublicAffairs

Chapter Ten

J. Jackson, *You Are What You Read* (2019), Unbound
Gapminder (data mapping project): https://gapminder.org

Part Four – Chapter Eleven

L. Hyde, *The Gift* (2012), Canongate Canons

M. Mauss, *The Gift* (2001), Routledge Classics

G. Vaughan, *For-Giving* (1997), Plain View, and free online version: http://www.gift-economy.com/forgiving.html

C. Eisenstein, *The More Beautiful World Our Hearts Know Is Possible* (2013), North Atlantic Books

R. Solnit, *Hope in Dark Places* (2010), Penguin

Impossible community open-source code: community. impossible.com

Chapter Twelve

J. M. Keynes, 'Economic Possibilities for Our Grandchildren', *Essays in Persuasion* (1932), Harcourt Brace: http://www.econ.yale.edu/smith/econ116a/keynes1.pdf

D. Pilling, *The Growth Delusion* (2018), Bloomsbury

Dr M. L. King, *Where Do We Go from Here? Chaos of Community* (2010), Beacon Press

R. Bregman, *Utopia for Realists* (2018), Bloomsbury

D. Graeber, *Bullsh*t Jobs* (2018), Penguin, and https://www.strike.coop/bullshit-jobs/

G. Standing, *Plunder of the Commons: A Manifesto for Sharing Public Wealth* (2019), Pelican

D. Elgin, *Voluntary Simplicity* (2010), HarperPaperbacks, second edition

To follow the Stockton UBI trial: https://seed.sworps.tennessee.edu/

Chapter Thirteen

M. Nelson, *The Argonauts* (2016), Melville House

R. Eisler, *The Chalice and the Blade* (1998), HarperCollins

Y. Gyasi, *Homegoing* (2016), Penguin

G. Lerner, *The Creation of Patriarchy* (1987), Oxford University Press

C. Merchant, *The Death of Nature* (1990), Bravo

A. Lorde, *I Am Your Sister* (2009), Oxford University Press

V. Woolf, *A Room of One's Own* (1929), Hogarth Press

P. Draper, '!Kung Women: Contrasts in Sexual Egalitarianism in Foraging and Sedentary Contexts' (1975): http://digitalcommons. unl.edu/cgi/viewcontent.cgi?article=1044&context=anthropol ogyfacpub

b. hooks, *Teaching Community* (2003), Routledge, and 'I Believe Whole-Heartedly That the Only Way out of Domination Is Love': https://opinionator.blogs.nytimes.com/2015/12/10/ bell-hooks-buddhism-the-beats-and-loving-blackness/

Chapter Fourteen

Y. H. Harari, *Sapiens: A Brief History of Humankind* (2015), Vintage

K. W. Kimmerer, *Braiding Sweetgrass* (2013), Milkweed Editions

M. K. Anderson, *Tending the Wild* (2013), University of California Press

M. Margolin, *The Ohlone Way* (1978), Heyday

D. Elgin, *Promise Ahead* (2001), Quill

D. Jensen, *The Myth of Human Supremacy* (2016), Seven Stories Press

M. Pollan, *How to Change Your Mind* (2019), Penguin

A. Grose, *A Guide to Eco-Anxiety* (2020), Watkins Publishing

Epilogue

C. Rovelli, *Seven Brief Lessons on Physics* (2015), Penguin

Index of Names

Huberts, Don: 'The end of the
    Oil Age' 84
Hughes, Chris 113, 117, 119, 345
Hyde, Lewis: *The Gift* 307, 310,
    368
Hyperloop 152, 169–72

Ice Watch 1, 14–15
Impossible 54, 112–13, 116, 310,
    311–22, 323, 325, 345
Impossible Foods 53–7, 59
Impossible to Print 316
Inglis, Bob 261
'Innovation in Giving' grant 317
Instagram 113–14, 117, 119, 124
International Criminal Court
    (ICC) 243
International Energy Agency
    (IEA) 140
International Institute for
    Sustainable Development
    (IISD) 146–7
International Monetary Fund
    (IMF) 258, 267
Iraq war protests (2003) 222
Iroquois 391

Jackson, Jodie 281
Jackson, Lisa 106, 107, 110, 111–12,
    131, 181, 258, 259, 263
Jefferson, Thomas 82
Joan of Arc 45
Jobs, Steve 101
Johnson, Lyndon B. 188
Johnson Publishing Company
    (JPC) 89–90
JP Morgan Chase 270
Juliana v. United States 240, 241

Keith, David 190; *A Case for
    Climate Engineering* 188

Keynes, John Maynard 332, 344; *The
    General Theory of Employment,
    Interest and Money* 346
Kickstarter 306
Kigali Accord (2016) 8–9
Kingelez, Bodys Isek: *U.N.*
    (cardboard sculpture) 284, 285
Kingsnorth, Paul: *Confessions of a
    Recovering Environmentalist* 192
Kintisch, Eli: *Hack the Planet* 189
Klein, Naomi x, 127, 130, 230, 244,
    379; *This Changes Everything*
    254–5
Knepp Castle 66, 67
Kobach, Kris: *The Referendum: Direct
    Democracy in Switzerland* 276
Kondo, Marie 338
Kopp, Sébastien 207–8
Kosonen, Kaisa 234
Kundalini yoga 401–2
!Kung 299, 364–7
Kuznets, Simon 330

Lackner, Klaus 180
Lane, Izzy 35
Langclaan, Jorne 161
Lantink, Duran 35
Lao Tzu 338
Larasi, Marai 376, 380, 382, 383
Laughlin, Robert 133
Lawrence, D. H.: 'Song of a Man
    Who Has Come Through' 306
Lechevallier, Aurélien 83, 232–3,
    256, 262, 264–5, 266
Lec, Jeffrey 142–3, 369
Lehman, Milton: *Robert H.
    Goddard: Pioneer of Space
    Research* 172
Leopold, Aldo 396–7
Lerner, Gerda: *The Creation of
    Patriarchy* 360, 367–8, 381, 383